been diagnosed
cer o
en c
oc he

a survey of what we know about breast cancer, but a personal search for guidance about navigating the complex and at times frightening terrain of breast cancer diagnosis, treatment, and survival.

Mary Ann G. Cutter is Professor in the Department of Philosophy at the University of Colorado, Colorado Springs. She holds a Ph.D. in Philosophy from Georgetown University through the Kennedy Institute of Ethics Program. She is the author of numerous publications on the philosophy of disease and bioethical topics, including AIDS, genetics, death and dying, and women's health care. In 1997, Professor Cutter received a CU system-wide service award for her work in developing genetic protection legislation for the state of Colorado.

THINKING THROUGH BREAST CANCER

THINKING THROUGH BREAST CANCER

A Philosophical Exploration of Diagnosis, Treatment, and Survival

Mary Ann G. Cutter

OXFORD
UNIVERSITY PRESS

OXFORD
UNIVERSITY PRESS

Oxford University Press is a department of the University of Oxford. It furthers the University's objective of excellence in research, scholarship, and education by publishing worldwide. Oxford is a registered trade mark of Oxford University Press in the UK and certain other countries.

Published in the United States of America by Oxford University Press
198 Madison Avenue, New York, NY 10016, United States of America.

Library of Congress Cataloging-in-Publication Data
Names: Cutter, Mary Ann Gardell, author.
Title: Thinking through breast cancer : a philosophical exploration of diagnosis, treatment, and survival / Mary Ann G. Cutter.
Description: New York, NY : Oxford University Press, [2018] |
Includes bibliographical references and index.
Identifiers: LCCN 2017033644 (print) | LCCN 2017040139 (ebook) |
ISBN 9780190637064 (online course) | ISBN 9780190637040 (updf) |
ISBN 9780190637057 (epub) | ISBN 9780190637033 (cloth : alk. paper)
Subjects: LCSH: Breast—Cancer—Diagnosis. | Breast—Cancer—Treatment. |
Breast—Cancer—Patients.
Classification: LCC RC280.B8 (ebook) | LCC RC280.B8 C83 2018 (print) |
DDC 616.99/449—dc23
LC record available at https://lccn.loc.gov/2017033644

1 3 5 7 9 8 6 4 2

Printed by Sheridan Books, Inc., United States of America

I dedicate this project to Vera Silvestro Gardell, Theresa Cutter, Anne Sarno, Faith Hansen, Isabelle Yurevitch, Lorraine Arangno, Jennifer Chamberlain, Lydia Cutter, Carol Gardell, Laura Christensen, and Myriam Miller, eleven great women who have spent hours upon hours listening to me rehearse the details of my clinical appointments and suggesting ways to think about the options that were before me. Their advice is captured here, albeit within the framework of philosophical language and my own limited musings.

"What is breast cancer?" has been a recurrent, central, if often unarticulated, question just below the surface of so many controversies about cause, prevention, treatment, prognosis, and policy. It also lies just below the surface of many individuals' difficult decisions.

—Robert A. Aronowitz, *Unnatural History: Breast Cancer and American Society*

CONTENTS

Preface *xiii*

1. Introduction 1
   Getting Started 1
   Why Think About Breast Cancer? 1
   Breast Cancer as a Case Study in Philosophy of Medicine 12
   Outline of Inquiry 15

2. How Is Breast Cancer Described? 18
   Personal Musings 18
   Realism 19
   Physicalism 28
   Reductionism 39
   Closing 46

3. How Is Breast Cancer Explained? 47
   Personal Musings 47
   Empiricism 49

Causal Relations 60
Cognitive Certainty 66
Cognitive Skepticism 73
Closing 77

4. How Is Breast Cancer Evaluated? 79
Personal Musings 79
Value Neutralism 81
Clinical Values 88
Value Certainty 95
Value Skepticism 100
Closing 102

5. How Is Breast Cancer a Social Phenomenon? 103
Personal Musings 103
Collaboration 105
Medicalization 113
Commodification 118
Politicalization 123
Closing 127

6. What Is the Relation Among the Descriptive, Explanatory, Evaluative, and Social Dimensions of Breast Cancer? 129
Personal Musings 129
An Interplay 130
Breast Cancer as an Integrative Medical Phenomenon 137
Breast Cancer as an Intersectional and Contextual Phenomenon 142
Breast Cancer as a Personalized Medical Phenomenon 146
Closing 149

7. What Are the Ethical Implications of Understanding and Treating Breast Cancer? 151
Personal Musings 151
Informed Consent and Autonomy 152
Risk Assessment, Nonmaleficence, and Beneficence 161
Access to Breast Cancer Care and Justice 171
An Integrative Ethical Approach to Understanding Breast Cancer 177
Closing 183

8. Extended Musings 185
Opening 185
Seven Lessons 186
Closing 196

Glossary of Medical Terms *197*
Glossary of Philosophical Terms *203*
References *207*
Index *221*

PREFACE

In late 2012, I was diagnosed with invasive ductal carcinoma of the right breast with nodal involvement and underwent a bilateral mastectomy, lymphectomy, and breast reconstruction. My experience as a breast cancer patient since late 2012 has included a number of notable clinical events: a screening mammogram, diagnostic mammograms and ultrasounds, a precancer diagnosis of ductal carcinoma in situ (DCIS), a heightened sense of risk, struggle with statistics regarding recommended interventions, a stereotactic (needle) biopsy that gave no new information, a surgical biopsy that disclosed DCIS as well as invasive ductal carcinoma, a cancer diagnosis, deliberation over surgical options, a bilateral mastectomy, a sentinel node biopsy, bilateral breast reconstruction, a lymphectomy, more struggle with statistics regarding chemotherapeutic and radiation options, still more struggle with statistics regarding adjuvant therapies, breast reconstruction revisions, and biannual blood and biomarker tests. Such was my entrance into becoming a breast cancer patient and a "citizen" of the "kingdom of the sick" (Sontag 1978, 3).

As a breast cancer patient, not only did I have numerous questions about my diagnosis, prognosis, and treatment, many of which were answered during and after my numerous clinical appointments, I found myself, as an academic philosopher of medicine, pondering philosophical questions about the nature of my breast cancer. I thought to myself that I cannot be the only person asking such questions, and so I penned this work. What follows is an inquiry into these philosophical questions, with a particular focus on the question "What is breast cancer?" I was inspired to pursue this question after reading the book, *Unnatural History: Breast Cancer and American History,* by University of Pennsylvania physician and historian of medicine Robert Aronowitz. I agree with Aronowitz when he says that the question "What is breast cancer?" lies "just below the surface of so many controversies about cause, prevention, treatment, prognosis, and policy" (Aronowitz 2007, 7). It surely did for me as I went about trying to navigate the uncertainties of my breast cancer diagnosis, prognosis, and treatment.

An example that illustrates the centrality of the question "What is breast cancer?" is illustrated in the following scenario. During one of my clinical appointments, I presented to one of my breast cancer doctors a report from the National Institutes of Health that discussed survival rates of breast cancer patients of certain age groups and diagnoses and who had undergone bilateral mastectomies followed by treatment. I identified with such patients and wanted to know more. Per the report, the survival rate was not 100%, and so I asked my doctor, "In what category will I fall, the survivor or the deceased?" The response (at least the one I remember) was "How can I say? I don't have a crystal ball!" Ah, herein lay the crux of an issue that still haunts me today: if doctors do not know the future prognosis of a clinical problem, how can they go about recommending treatment? How can physicians recommend to a breast cancer patient that she be cut, burned, and poisoned when knowledge of breast cancer is not more certain? Put more personally, why did I go through such harsh surgical treatments if my breast cancer is not fully understood, the etiology is not clear, the future cannot be predicted, and the treatment plan is based on incomplete knowledge of breast cancer? How and why did clinicians underestimate the extent of my breast cancer in the initial stages of diagnosis given the current screening and diagnostic technologies that are available? How and why did a clinician estimate my risk level at "50/50" for breast cancer, when soon after I was diagnosed with invasive ductal carcinoma? But this is the past. So the more productive question is this: given what we know about breast cancer, how can I or any other breast cancer patient navigate the uncertainties of breast cancer diagnosis, prognosis, and treatment? This is a central concern I set out to explore in this inquiry.

In what follows, I subdivide the question "What is breast cancer?" into six questions that help tease out issues concerning the diagnosis, prognosis, and treatment of breast cancer and submit them to philosophical analysis. These questions include (1) How is breast cancer described?, (2) How is breast cancer explained?, (3) How is breast cancer evaluated?, (4) How is breast cancer a social phenomenon?, (5) What is the relation among the descriptive, explanatory, evaluative, and social dimensions of breast cancer?, and (6) What are the ethical implications of how medical practitioners understand and treat breast cancer? In addressing each of the questions, the inquiry that follows offers a reconstruction of the descriptive, explanatory, evaluative, and social dimensions of breast cancer. It illustrates how the descriptive, explanatory, evaluative, and social dimensions of breast cancer are integrated. That is, the descriptive, explanatory, evaluative, and social dimensions of breast cancer mutually define and situate each other, leading to an integrative account of breast cancer. An integrative account of breast cancer carries ethical implications for how informed consent is secured, risk assessment is conveyed, and social justice is achieved in breast cancer care. On this account, breast cancer is an evolving notion that provides structure, significance, and social context to clinical reality. It is in this framework that uncertainty is lodged and is appropriately addressed.

Readers from a wide variety of backgrounds may find the analysis of interest to them. These include philosophers of medicine and bioethicists as well as clinicians in breast cancer medicine. These include as well both graduate and undergraduate students in courses in philosophy of medicine, bioethics, feminism, and women's health, as well as those who gather in book clubs and have an interest in women's health issues. I also hope to reach out to breast cancer patients who have asked the kinds of questions I have had about breast cancer and who have not found discussions in the mainstream breast cancer literature helpful. In this spirit, I have tried to make this analysis neither overly clinical nor overly academic, although, when I do raise a clinical or academic matter, I offer definitions and sources for those interested in pursuing particular lines of thought. I recognize that my audience is broad and, as a consequence, I recognize that the level of clinical discussion or philosophical discussion may not reach a level of sophistication that a reader may wish. Such a level is not my intent here; my intent is to show what reflections at the intersection of medicine and philosophy look like, especially from the perspective of a first-person narrative and a desire to communicate such a perspective in ways that may make sense to others who have asked similar kinds of questions about breast cancer.

The clinical literature used in this analysis is considered widely accepted, allopathic or biomedical literature on breast cancer from the American Cancer Society, the US National Cancer Institute of the National Institutes of Health, and mainstream Western medical journals and books. By "allopathic," I mean the medical use of surgical and pharmacological interventions to treat or suppress symptoms or pathophysiological processes of diseases and other clinical conditions based on clinical evidence. I focus primarily on this literature for three reasons. First, I was treated by allopathic doctors and was encouraged to read the allopathic literature and so draw from what I have learned during my experience as a breast cancer patient. Second, my interest is in submitting the current allopathic view of breast cancer to philosophical analysis in order to deconstruct its assumptions and claims. Third, the current allopathic view of breast cancer provides sufficient material for a philosopher to analyze without drawing upon "unorthodox," "holistic," "alternative," "complementary," or "traditional folk" views about the diagnosis and treatment of breast cancer, although one will see that I end up extending the allopathic view of breast cancer to include views drawn from some of the aforementioned traditions in medicine.

The philosophical literature used in this analysis is drawn from major figures in the field of Western philosophy. The term "philosophy" comes from the Greek roots *philo*, meaning "loving," and *sophia*, meaning "wisdom." As a discipline that pursues the love of wisdom, philosophy in the West has a long history dating back to the Ancients such as the Greek philosophers Plato (428–348 B.C.E.) and Aristotle (384–322 B.C.E.). As philosophers like to say, philosophers ask "questions that matter" (Miller, 1987), such as "What is reality?," "How do we know reality?," and "How ought I (and we) to be and do in light of my (and our) view of reality and knowledge?" Such questions correlate with particular areas of study in philosophy, including ontology, epistemology, and ethics, respectively. Here, ontology refers to the study of being, epistemology to the study of knowledge, and ethics to the study of values. In answering ontological, epistemological, and ethical questions, philosophers typically seek to provide a systematic analysis of a subject matter.

Beyond the general categories of ontology, epistemology, and ethics, there are numerous areas of foci in philosophy. An area that provides a fair amount of the resources that I cite in this inquiry is philosophy of medicine. Here, philosophy of medicine is a field in philosophy that brings together a range of philosophical investigations of topics raised in and by medicine. Here medicine is understood broadly to include the discipline that studies and treats problems that patients bring to clinicians' attention. Philosophy of medicine has a long history dating back to the early clinicians, such as Hippocrates (c. 460–c. 370 B.C.E.) and Galen (c. 130–200 B.C.E.), each of whom developed an influential account of health and disease that has endured well into the modern period of medicine. Modern clinicians, such as English physician Thomas Sydenham (1624–1689) and Scottish physician William Cullen (1710–1790), called attention to the need to ground clinical diagnostic and treatment claims on empirical observations. Others, such as French anatomist and physiologist Xavier Bichat (1771–1802) and German physician and pathologist Rudolf Virchow (1821–1902), emphasized the need to investigate disease and its treatment through physiological and pathological frameworks. Today, philosophers of medicine, such as H. Tristram Engelhardt, Jr. (1941–present), Susan Sherwin (1947–present), and Rosemarie Tong, contribute to discussions regarding the nature, purpose, and ethical standards of medicine. Such literature continues to grow and gives us the opportunity for philosophical reflection on questions that matter in medicine.

The focus of this inquiry is on answers to the question "What is breast cancer?" for purposes of shedding light on how to deal with the uncertainty of breast cancer, diagnosis, prognosis, and treatment. While I think the inquiry and its conclusions find great relevancy in discussions of the diagnosis, prognosis, and treatment of other clinical conditions, I will not wander too far from the topic of breast cancer in the reflections and analyses offered here. An inquiry into the particularities and uncertainties of diagnosis, prognosis, and treatment of other clinical conditions would be, in my view, the focus of another project. Nevertheless, I do not discourage readers from drawing their own lessons from this analysis for other clinical conditions that they have in mind. I have no doubt that reflections on the uncertainty of breast cancer diagnosis, prognosis, and treatment can lead to reflections on how to manage the uncertainty of knowledge regarding other clinical conditions.

While I have been writing in the philosophy of medicine and biomedical ethics for some time now, I have never taken on such a personal project. I decided to pursue this project because, over these past years as a breast cancer patient, I have been struck with the relevancy of my philosophical training in helping me come to terms with the uncertainty and ambiguities of the diagnosis, prognosis, and treatment of my breast cancer. Alternatively, as a philosopher of medicine, I have been struck with how breast cancer provides an opportunity to think through long-standing philosophical issues about the nature of reality, knowledge, and values. Such a relation between medicine and philosophy is not new. As philosopher Michel Foucault (1926–1984) tells us, "The clinician's description, like the philosopher's analysis, proffers what is given by the natural relation between the operation of consciousness and the sign [of disease]" (1973 [1963], 95). In other words, observations of clinical reality take place through the operations of consciousness or lenses we bring to bear on observations of clinical reality. So here goes, not a diary, but an immersion into thinking philosophically about breast cancer, taking us back to our roots of thinking in medicine when clinicians entertained philosophical questions about conditions that present in the clinic. My hope is that what follows is not taken as an edict about how breast cancer patients ought to think and decide about their care. Such is not my intent. What follows is offered as an example of how I have tried to come to terms with managing the uncertainty of my breast cancer diagnosis, prognosis, and treatment.

A project such as this could not have been completed without others. First and foremost, I thank those in medicine who have cared and are caring for me. I hope that my reflections in no way indicate that my care has been less than excellent and less than deeply appreciated. I am grateful to Dr. Leslie Ahlmeyer of Yampa Valley OB/GYN in Steamboat Springs, Colorado, for being the first to raise concern about a lump in my right breast. Thanks go to Dr. Fred Jones and Dr. Robert Lile, radiologists at Yampa Valley Medical Center in Steamboat Springs, Colorado, who encouraged me to pursue more tests, despite there being inconclusive evidence of breast cancer. I am grateful to Dee Peterson, Patient Navigator at Penrose Hospital's Center for Women's Imaging in Colorado Springs, Colorado, who recommended, despite the appearance of lack of urgency in my mammogram results, that I see a breast care specialist. I am so very grateful to Dr. Toni Green-Cheatwood, Director of Breast Oncology for the Southern Colorado Breast Care Specialists in Colorado Springs, Colorado, for following her "gut instinct" in removing an area of my right breast that was not the focus of concern on my mammograms. It was this area that turned out to be invasive ductal carcinoma of the breast with nodal involvement and that led to more aggressive clinical interventions. I am so very grateful to Dr. Aaron D. Smith, a plastic and reconstructive surgeon in Colorado Springs, Colorado, for joining Dr. Green-Cheatwood's team and reconstructing my breasts, something that I never thought I would be in the position of considering given my long-standing feminist view that women should not be defined by their body parts. My appreciation of Dr. Smith's clinical and artistic talents in reconstructive surgery is immense. I thank Dr. Hyun Sue Kim, medical oncologist at Rocky Mountain Cancer Center in Colorado Springs, Colorado, for carefully and diligently overseeing my oncology needs and answering my numerous questions every time I see her.

I thank all those who have continued to be by my side on this journey. My family and friends never paused in their assistance, support, and love. In particular, I thank my husband, Lew, and our beloved children, Lewis, Theresa, and John, for their unconditional support during a most challenging time in my life. Thank you for all the times you listened and listened and listened! I also thank my dad, John P. Gardell, Sr. (who passed away in July 2014), as well as Paige Holy Cutter, Christine Nolan Cutter, Robyn Canales, Liz Cutter, Steve Cutter, John Gardell, Jr., Steve Gardell, Matt Gardell, Pete Gardell, Sophie Melvin, Betsy Johnston, Sheryl Botts, Aleisha County, Kevin Abney, Fran Pilch, Jody Proper, Kristen Miller, Dave Miller, Tommy and Annmarie Ryan, Debra Bergoffen, George Schroeder, Rita Hug, Ellen Race, Raphaela DiBauda, and Rita Simon Soller for their willingness to step into my shoes when lending advice. My colleagues in the Department of Philosophy at the University of Colorado at Colorado Springs never ceased to be supportive, which means so very much since my initial surgeries occurred in the middle of a busy and active semester while I was Chair of the department. I am indebted to Raphael Sassower, Ian Smith, Dorothea Olkowski, Rex Welshon, Jeff Scholes, Sonja Tanner, Mary Jane Sullivan, Allison Postell, Jennifer Jensen, Fred Bender, and Monica Killebrew for their willingness to do whatever I needed in order to keep the department running and my health at the forefront of my daily life. I am thankful to Dean Peter Braza and Senior Associate Vice Chancellor David Moon for their steadfast support during a time when even the smallest tasks seemed insurmountable. I remain indebted to H. Tristram Engelhardt, Jr., for his insightful work on philosophy of disease and for the encouragement across the miles to continue work in this area of philosophy of medicine. I also remain indebted to Laura Sansone Keyes, who shared with me many years ago her own struggle with breast cancer prior to losing her life to it. I dedicate this project to Vera Silvestro Gardell, Theresa Cutter, Anne Sarno, Faith Hansen, Isabelle Yurevitch, Lorraine Arangno, Jennifer Chamberlain, Lydia Cutter, Carol Gardell, Laura Christensen, and Myriam Miller. May this dedication be seen as a humble way to express my sincere thanks to these great women for being by my side. And, as always, the spirit of my mother, M. Theresa Gardell, continues to inspire me in all my academic projects, and especially this one on cancer, a condition that affected her lungs and that took her life in 1985 while I was a graduate student in philosophy studying philosophy of medicine at Georgetown University in Washington, D.C.

Colorado Springs and Steamboat Springs, Colorado
September 20, 2017

THINKING THROUGH BREAST CANCER

Chapter 1

Introduction

GETTING STARTED

The project that follows investigates how philosophical inquiry can help us understand breast cancer and how to manage the uncertainty of breast cancer diagnosis, prognosis, and treatment. In keeping with this, let us begin with a consideration of why we should think about how medical practitioners and researchers understand breast cancer, how breast cancer serves as a case study in the philosophy of medicine, and how this project is organized.

WHY THINK ABOUT BREAST CANCER?

There are a number of reasons to reflect on how breast cancer is understood in Western medicine. To begin with, breast cancer is a prominent nosology or clinical classification in medicine. What and how doctors diagnose guides what actions are advised to be taken and which ones are not, as well as who is charged with what tasks over what recommended period of time. If one is diagnosed with Stage IA cancer, for instance, there will be certain avenues of treatment that will be recommended, such as a lumpectomy and radiation. If one is diagnosed with Stage III breast cancer, there will be other avenues of treatment that will be recommended, such as a mastectomy with chemotherapy and radiation. In each scenario, specific diagnoses call for certain treatments, overseen by certain types of specialists, including radiologists, breast cancer surgeons, pathologists, oncologists, reconstructive surgeons, specialized nursing staff, psychologists, and social workers (National Cancer Institute 2014d, Taghian et al. 2009).

A second reason to study breast cancer is that it affects a large number of women in the United States and around the world. Breast cancer is one of the most common types of cancer among women in the United States (National Cancer Institute 2015f). Today, women in the United States are said to have a one in eight (or 12%) chance of having breast cancer (American Cancer Society 2015e, 1). According to the National Cancer Institute at the US National Institutes of Health, it is estimated that, in 2012, 226,870 women and 2,190 men were diagnosed with breast cancer, and 39,510 women and 410 men died of the condition (National Cancer Institute 2014a; "Breast Cancer in Men," 2014). In this way, breast cancer is the second leading cause of death among women in the United States, following heart disease. It is the second leading cause of death to cancer among women in the United States, following lung cancer (American Cancer Society 2015e). Globally, according to the World Health Organization, breast cancer is the most common cancer in women both in developed and less developed countries and affects approximately 1.7 million women worldwide. As it reports, "[i]t is estimated that worldwide over 508,000 women died in 2011 due to breast cancer. Although breast cancer is thought to be a disease of the developed world, almost 50% of breast cancer cases and 58% of deaths occur in less developed countries" (World Health Organization 2014, ¶1).

In the United States, the incidence rates of breast cancer have "been stable over the last ten years" (National Cancer Institute 2015f, 7). The American Cancer Society reports a 7% drop in incidence rates of breast cancer from 2002 to 2003 following the release of results from the Women's Health Initiative study (Women's Health Initiative 2015) and the decline in the use of therapeutic hormones among women (American Cancer Society 2015e). In terms of mortality from breast cancer, "[d]eath rates have been falling on average 1.9% each year over 2002–2011" (National Cancer Institute 2015f, 7). Worldwide, the incidence rates of breast cancer in developing countries are increasing, but at a rate that is less than in developed countries. In developing countries, an increase in the incident rates of breast cancer is associated with longer lifespans and the effects of urbanization on diet and lifestyle (World Health Organization 2014, 2015).

Of growing interest is that breast cancer affects women of particular ethnic backgrounds in different ways. According to the National Cancer Institute's Surveillance, Epidemiology, and End Results Program, in 2007 through 2011, the incidence rate for breast cancer for whites was 128 per 100,000 women. For blacks, it was 122.8; Asians/Pacific Islanders, 93.6; American Indians/Alaska Natives, 79.3; Hispanics, 91.3; and non-Hispanics, 129.7. In 2007 through 2011, the death rate for breast cancer for whites was 21.7 per 100,000 women. For blacks, it was 30.6; Asians/Pacific Islanders, 11.3; American Indians/Alaska Natives, 15.2; Hispanics, 14.5; and non-Hispanics, 22.9 (National Cancer Institute, 2015f, 1–2). On a global scale, and according to the World Health Organization, "[i]ncidence rates vary greatly worldwide from 19.3 per 100,000 women in Eastern Africa to 89.7 per 100,000 women in Western Europe. In most of the developing regions, the incidence rates are below 40 per 100,000. The lowest incidence rates are found in most African countries, but here breast cancer incidence rates are also increasing" (2014, ¶2). Survival rates for breast cancer vary greatly, "ranging from 80% or over in North America, Sweden and Japan to around 60% in middle-income countries and below 40% in low-income countries. The low survival rates in less developed countries can be explained mainly by the lack of early detection programs, resulting in a high proportion of women presenting with late-stage disease, as well as by the lack of adequate diagnosis and treatment facilities" (World Health Organization 2014, ¶3).

A third reason to study breast cancer is that, in the United States, significant funds are spent on breast cancer. According to one study, $16.50 billion was spent on breast cancer care in the United States in 2010 (National Cancer Institute 2015b). Compare this to $14.14 billion on colorectal cancer, $12.12 billion on lung cancer, $11.85 billion on prostate cancer, $5.44 billion on leukemia, $5.12 billion on ovarian cancer, and $2.36 billion on melanoma (National Cancer Institute 2015b). Estimates are that in 2020, $20.50 billion will be spent on breast cancer, and a record amount compared to other types of cancer (National Cancer Institute 2015b). With regard to research, the National Cancer Institute of the National Institutes of Health spent $602.7 million on breast cancer research in the United States in 2012 (National Cancer Institute 2014b). Compare this to $314 million for lung cancer, $265 million for prostate cancer, $256 million for colorectal cancer, $234 million for leukemia, $121 million for melanoma, and $111 million for ovarian cancer. The bottom line is that significant funds are spent on breast cancer care and research (National Cancer Institute 2015b).

A fourth reason to study breast cancer is that breast cancer is the focus of numerous laws and policies. In the United States, the Breast Cancer Detection and Demonstration Projects of the 1970s marketed new public health practices such as regular screening mammograms and physical examination of the breast to community groups, medical specialists, and local hospitals (Aronowitz 2007, 237). Under the Women's Health and Cancer Act of 1998, group health plans, insurance companies, and health maintenance organizations

(HMOs) must provide coverage for services relating to a mastectomy in a manner determined in consultation with the attending physician (US Department of Labor 2013). This includes all stages of breast reconstruction, external breast prostheses, and treatments of the physical complications of the mastectomy, including lymphedema (i.e., a condition marked by localized fluid retention and tissue swelling). According to the 1990 Americans with Disabilities Act (ADA), those with cancer and those who have had cancer are considered to have a disability (American Cancer Society 2008). A "disability" applies if an individual has a physical or mental problem that substantially limits one or more of her or his "major life activities," one has a record of having such a problem in the past, and other people recognize that one has a problem. Some of the "major life activities" that are covered include, but are not limited to, caring for oneself, doing manual tasks, seeing, hearing, eating, sleeping, walking, standing, lifting, bending, speaking, breathing, learning, reading, concentrating, thinking, communicating, and working. On January 1, 2009, the ADA Amendments Act of 2008 went into effect and made some major changes to the way the definition of "disability" has been interpreted by the ADA in the past. The 2009 Amendments cover major bodily functions, including, but not limited to, functions of the immune system, normal cell growth, bladder, brain, nervous system, and reproductive system, among others. These changes were applied to people with cancer, including those with breast cancer (American Cancer Society 2008).

The Breast and Cervical Cancer Prevention and Treatment Act of 2000 (Public Law 106-354) (Centers for Disease Control 2013) gives states the option to provide medical assistance through Medicaid for eligible women who are found to have breast or cervical cancer, including precancerous conditions. The Native American Breast and Cervical Cancer Treatment Technical Amendment Act of 2001 (Public Law No. 107-121) (Centers for Disease Control

2013) amends Title XIX of the Social Security Act to clarify that Indian women with breast or cervical cancer are eligible for health services provided under the Indian Health Service or a Medicaid eligibility option for tribal organizations. The Affordable Care Act was signed into law on March 23, 2010, and it helps make prevention affordable and accessible to all Americans by requiring most health plans to cover recommended preventive services while cost sharing. In the case of breast cancer, the Act covers preventive care services, such as mammograms, and eliminates the costs to patients for these proven services in all new health plans. It also eliminates co-pays and deductibles for preventive services under Medicare, such as mammograms ("The Affordable Care Act" 2013). (As of this writing, the Affordable Care Act may be revised by the new Trump Administration.)

In addition, there are numerous US state laws that address a number of areas relating to breast cancer. Most states in the United States mandate coverage for a mastectomy, prosthetics, and breast reconstruction, as well as mammogram screenings. Other states (e.g., Colorado) require breast cancer care facilities to be accredited, and others (e.g., California) require clinicians to inform patients of alternative therapies. Still others (e.g., New York) permit an income tax write-off for breast cancer funds, and others (e.g., Texas) regulate the length of stay for inpatient care following a mastectomy (US Department of Health and Human Services 2013).

Outside the United States, the World Health Organization (WHO) is concerned about breast cancer because the incidence of breast cancer is increasing in the developing world due to an increase in life expectancy and the effects of urbanization on diet and lifestyle. Recall the data that were shared earlier. The incidence rate of breast cancer in eastern Africa is 19.3 per 100,000 women and 89.7 per 100,000 women in western Europe. Yet the survival rate for breast cancer approximates 80% or over in developed countries such as found in western Europe compared to 40% in low-income countries, such as found in eastern Africa. Given this situation, WHO says that "[e]arly detection in order to improve breast cancer outcome and survival remains the cornerstone of breast cancer control" (World Health Organization 2014, ¶7). It encourages low-cost screening approaches, such as clinical breast exams, in limited resource settings and the introduction of more advanced screening methods, whenever possible. WHO also encourages a closer look at the following question: "What is it about women's lifestyles in developed countries that makes them so much more likely to develop the disease than their counterparts in parts of southwest Asia and Africa where incidence is typically five times lower?" (World Health Organization 2015, ¶6).

Globally speaking, policies, laws, and practices that focus on breast cancer vary (National Cancer Institute 2015c). With regard to breast cancer screening, breast cancer screening is covered annually under the US Affordable Care Act of 2010 for women aged 40 and above ("The Affordable Care Act" 2013). In Canada, breast cancer screening is covered every two years for women 40 and over (Canadian Cancer Society 2007). In Uruguay, a national program funds breast cancer screening for women 40 through 69 every two years (National Cancer Institute 2015c). In the Netherlands, the Breast Cancer Screening Programme offers annual breast cancer screening to women 50 through 75 ("Harms Outweigh" 2014). In Australia, the National Breast Screening Program offers breast cancer screenings to women aged 50 through 69 every two years (National Cancer Institute 2015c). In Singapore, BreastScreen Singapore publicly funds half of the cost of breast cancer screening for women 50 through 64 every two years (Health Promotion Board 2015). In South Korea, a national program offers breast cancer screening to women 40 through 75 every two years (National Cancer Institute 2015c). In China, breast cancer screening is advised for women 40 through 59 every three years (National Cancer Institute 2015c). As seen here, breast cancer is the focus of numerous laws and policies around the globe.

A fifth reason to study breast cancer is that our understanding of breast cancer provides a glimpse into the history of understanding breast cancer. According to scholars, Greek physician Hippocrates, the father of Western medicine, supplied the earliest description of breast cancer (Olson 2002, 12–13). He called cancer "*karkinos*," a Greek word for crab, because cancer tumors were seen to send out tentacles of sorts that "grab" or surround normal tissue. Cancer, he held, occurred because of an excess of one of the body's four humors, black bile. One might recall that Hippocrates held that the body had four humors (namely, black bile, yellow bile, phlegm, and blood), each corresponding to one of the traditional four temperaments (namely, irritability, aggression, calmness, and playfulness) and elements of the world (namely, earth, fire, water, and air) (Cutter 2003, 48). According to Hippocrates, the treatment of breast cancer focused on balancing the black bile in the body. But if left untreated, tumors in the breast became larger, harder to touch, and darker in appearance, pushing aside other tissues and eventually ulcerating through the skin and generating a harsh odor and dark fluid (Olson 2002, 12–13). If left untreated, tumors in the breast spread to other parts of the body, which resulted in the death of the patient.

Hippocrates' humoral account of breast cancer remained a dominant account well into the nineteenth century. It began to be revised with the development of new knowledge in oncology. Here "oncology" is from the Greek roots *oncos*, meaning "swelling" or "mass," and *logy*, meaning "the study of" (American Cancer Society 2015c). In the late nineteenth century, breast oncology began its development as we know it today. Clinicians held that breast cancer spread in a contagious, stepwise fashion from the primary tumor in the organ and tissue, and the only course of treatment was to remove the location of the cancer and everything in its path. In 1895, physician William Halstead pioneered the radical mastectomy at the Johns Hopkins Hospital, in Baltimore, Maryland, and the procedure became routine treatment for breast cancer well into the twentieth century (Olson 2002, 45). Here the "radical mastectomy" (sometimes called the "the Halstead mastectomy" or "super radical mastectomy") involved the removal of the breast tissue, underlying chest muscles, and axillary (i.e., under the arm) lymph nodes. In 1959, the American Joint Committee on Cancer (AJCC) developed the tumor, node, metastasis (TNM) staging system for breast cancer "in the absence of effective systematic therapy and certainly in a void of the understanding of the biology of breast cancer that exists today" (American Joint Committee on Cancer 2010, 422). The staging system involves an assessment of the size and location of the tumor, the extent of tumor involvement, and the degree to which breast cancer has spread. It continues to be used today.

In the early twenty-first century, much research is focused on the biomolecular nature of breast cancer. A biomolecular account of breast cancer involves an explanation of breast cancer based on its component parts, such as genes, biomarkers, and particular cells. Research is focused on the efficacy of adjuvant therapy (i.e., treatment given after the primary therapy to increase the chance of long-term survival), systematic therapy (i.e., therapy that targets metastatic cancer), alternative treatments to disfiguring surgery (e.g., the replacement of mastectomies with lumpectomies), and safer and more targeted radiation techniques and chemotherapeutic interventions (American Joint Committee on Cancer 2010, 422–423). Clinicians are currently calling for more research on immunotherapies for breast cancer and more attention to targeted pharmaceutical treatments for metastatic breast cancer ("Novel" 2014). In short, our understanding of breast cancer has changed since ancient times. Hippocrates' humoral account of breast cancer is replaced with ones based on organ- and tissue-based accounts, which are being replaced by biomolecular and genetic accounts (American Cancer Society 2015c).

A sixth reason to study breast cancer is that a clinical understanding of breast cancer continues to require clarification. We now know more about the genetic make-up of breast cancer, but we do not understand how the genetic mutations come about. We now know more about the biomolecular make-up of certain breast cancers, but we do not understand how to treat certain kinds of breast cancer any differently than other types. Medical practitioners and researchers now knows more about the role of lymph nodes in determining the prognosis of breast cancer, but we do not understand why some breast cancers progress very quickly and others do not. We know more about very early breast cancers but do not understand which very early breast cancers develop into late-stage breast cancers and which do not. Medicine supports a number of health policies that guide breast cancer care, such as mandatory coverage for mammograms, but we do not understand how to screen and test for breast cancer in ways that avoid false-positives (i.e., cases that are said to be breast cancer but are not) and false-negatives (i.e., cases that are said not to be breast cancer but are). In short, the current understanding of breast cancer is in need of reconsideration and revision.

Related to the need to clarify the current understanding of breast cancer is the need to investigate how patients and breast cancer specialists navigate the uncertainties of breast cancer diagnosis, prognosis, and treatment. Given the limited and evolving clinical information we have about breast cancer, it follows that patients as well as breast cancer specialists make decisions in light of information that is uncertain. The question arises, how does one make decisions about breast cancer diagnosis, prognosis, and treatment given that clinical information is not 100% certain? Why is such clinical information uncertain? What methods do we rely on to make decisions in a context of uncertainty and risk? What practical advice can be given about making such decisions? These are the kind of questions that focus on how to navigate the uncertainties of breast cancer diagnosis, prognosis, and treatment and receive little attention in the breast cancer literature.

A seventh reason to study breast cancer is that medicine's understanding of breast cancer carries ethical implications. As previously indicated, how breast cancer is understood sets the stage for how it is diagnosed and treated. Further, how breast cancer is diagnosed and treated provides the basis for the informed consent process and the ways in which patient autonomy is protected in the clinical setting. It provides the basis for determining how we are to minimize harms and maximize benefits for breast cancer patients and the ways in which the duties of nonmaleficence and beneficence are carried out in breast cancer care. It serves to guide decisions about the allocation of health care resources in breast cancer medicine, and to whom they are distributed and for what purposes. It frames the development of health laws and policies regarding breast cancer care at the local, regional, national, and global levels. How breast cancer is understood carries ethical implications for patient autonomy, the welfare of patients, and social justice in breast cancer medicine.

In short, breast cancer is a prominent nosology and nosography in medicine which affects a significant number of women, commands a large portion of financial resources, serves as the focus of numerous policies and laws, provides a glimpse into history, calls out for clarification, and carries ethical implications. There are a number of reasons to submit breast cancer to philosophical investigation.

BREAST CANCER AS A CASE STUDY IN PHILOSOPHY OF MEDICINE

In what follows, breast cancer serves as a case study in philosophy of medicine for purposes of asking questions about the nature of breast cancer, how we know it, and how we treat it. As a subdiscipline of philosophy, philosophy of medicine brings together a range of philosophical pursuits on topics raised in and by medicine (Pellegrino 1976, 1978; Engelhardt 1977, 2000; Engelhardt and Erde 1980, 365–366; Tong 1997; Sherwin 1998; Stempsey 2004; Marcum 2008). Here topics include ones found in the basic sciences (e.g., how normal cells function), theoretical endeavors (e.g., the development of diagnostic, explanatory, and therapeutic models in medicine), and actual practices (e.g., the diagnosis, prognosis, and treatment of breast cancer in the clinic). As a result, one can have philosophical questions about how structure, function, causality, and clinical decision making and treatment are understood in medicine. In the case of breast cancer, there are philosophical questions about the distinction between normal and abnormal cell function; what constitutes sound diagnostic, explanatory, and therapeutic models in breast cancer care; and what constitutes sound diagnostic, prognostic, and therapeutic practices in breast cancer medicine. As a result, one can have philosophical questions about the structure and function of the breast; the diagnosis, etiology, prognosis, and treatment of breast cancer; and clinical decision making and treatment in breast cancer medicine.

A major focus of philosophy of medicine, and one that is evident in this inquiry, is on disease (Caplan et al. 1981; Engelhardt 1984; Reznek, 1987; Engelhardt and Wildes 2003; Cutter 2003, 2012; Pellegrino 2004). As physician and philosopher of medicine H. Tristram Engelhardt, Jr. puts it, philosophy of medicine "offers promise of a special set of conceptual issues bearing on the status of concepts such as disease, illness, and health" (1977, 104). Such issues include the nature of the clinical event, how we know the clinical event, how we evaluate the clinical event, how we socialize the clinical event, how we understand the relations among such conceptual dimensions, and how we respond to the clinical event in light of its nature, knowledge, evaluation, and social framework. In the case of breast cancer, questions arise concerning its nature, how we know it, how we evaluate it, how we socially frame it, how we understand the relation among its dimensions, and how we treat it in light of our understanding and evaluation of the phenomenon in its social context.

The concepts of disease, illness, and health "have unique explanatory and evaluative uses" (1977, 104). Consider the difference between the scientific and clinical endeavors of knowing an object of investigation. In investigating stars, for instance, the scientist typically does not have any interest in manipulating them. In contrast, in understanding disease, the clinician has an interest in understanding it for purposes of treating that which patients bring to the attention of health care professionals. As Engelhardt says, the concept of disease "deals with suffering and one becomes interested in the physiological bases of diseases because of concern for the sufferings and debilities associated with illnesses" (Engelhardt 1977, 101). Granted, in this day and age of genetic medicine, the concept of disease increasingly addresses presymptomatic conditions detected in the clinical laboratory. Nevertheless, the motivation to treat presymptomatic conditions is similar to that of disease: treatment. If presymptomatic conditions are left alone, they may develop into clinical problems that may lead to pain, suffering, and perhaps death in the lives of patients. If presymptomatic conditions are treated, there is a chance that the pain, suffering, and perhaps death of the patient is minimized, if not all together prevented. In the end, disease involves biological dysfunction as well as the undermining of goals, wishes, or desires as experienced by patients.

According to physician and philosopher of medicine Lawrie Reznek, a philosophical investigation of disease has additional practical implications. It can "make a real contribution to settling disputes over disease classification" (1987, 11). In so doing, it lends support to understanding the concept of disease. In addition, "it enables individuals to justly claim medical insurance, to have the treatment of their illness covered by a national health service, to avoid having normal conditions subjected to medical treatment, to avoid being unjustly punished for the symptoms of an illness, and to enable us to decide whether the problem should be dealt with by some other institution" (Reznek 1987, 11). In other words, a philosophical study of disease helps clarify what conditions medicine can and ought to treat and which ones third parties ought to cover. In the case of breast cancer, a philosophical study of breast cancer helps clarify what medicine can and ought to treat and which breast conditions ought to be covered by third parties.

Philosophically speaking, then, disease is an expansive notion. It involves the interests of clinical and medical research professionals, patients, third parties, clinical and medical research professionals, health care institutions, and members of society in describing, explaining, evaluating, and socializing that which comes into the clinic. According to Engelhardt, these endeavors of describing, explaining, evaluating, and socializing disease reflect "four conceptual dimensions" or "modes of medicalization" (1996, 195) in medicine. They constitute the "language of medicine" (1996, 195) in that they provide the "grammar" and "rules," so to speak, for constructing meaning about and practical guidelines for addressing clinical problems. They constitute "four different clusters of 'syntactical' and 'semantical' constraints that shape the ways we speak of, understand, and experience medical reality" (Engelhardt 1996, 195). By "syntactical" Engelhardt meant of or relating to the rules of the grammatical arrangement of words. By "semantical," he meant of or relating to the meaning of words. The endeavors of describing, explaining, evaluating, and socializing disease involve "a set of descriptive assumptions" (Engelhardt 1996, 207), "causal explanatory models" (Engelhardt 1996, 196), "evaluative assumptions regarding which functions, pains, and deformities are normal in the sense of proper and acceptable" (Engelhardt 1996, 196), and "social expectations regarding individual ills or particular forms of sickness" (Engelhardt 1996, 196).

The inquiry that follows relies on Engelhardt's analysis of the four dimensions of disease as it investigates the nature of breast cancer, how we know it, and how we treat it within our social contexts. It extends Engelhardt's analysis by focusing more extensively on the interrelations between and among the multiple dimensions of disease as found in the case of breast cancer. In the end, the inquiry offers an integrative account of breast cancer that warrants a personalized medical approach, thus emphasizing the intersectionality among the dimensions of breast cancer. Personalized, or precision-based, medicine offers the ability to provide information on the individualistic characteristics of a breast cancer as well as its response to treatment ("What Is Personalized . . . ?" 2017). The inquiry further extends the analysis by illustrating the ethical implications of understanding breast cancer in an integrative way. It closes with practical suggestions regarding how to navigate the uncertainties of breast cancer diagnosis, prognosis, and treatment.

OUTLINE OF INQUIRY

In what follows, I submit the question "What is breast cancer?" to philosophical analysis. Generally put, each chapter begins with a series of personal musings as a breast cancer patient that have led me to pen this work. Each chapter then goes on to address such musings through the lens of philosophical discussion on particular clinical matters concerning our understanding of breast cancer. More specifically, Chapter 2 asks the question "How is breast cancer described?" Here the philosophical concepts involve matters of ontology, and specifically the philosophical problems of realism, materialism, and reductionism. Chapter 3 asks the question "How is breast cancer explained?" Here the philosophical concepts involve matters of epistemology, and specifically the philosophical problems of empiricism, causality, and cognitive certainty. Chapter 4 asks the question "How is breast cancer evaluated?" Here the philosophical concepts come from axiological discussions about how values shape breast cancer diagnosis, prognosis, and treatment. Specifically, the chapter probes the philosophical problems of value neutralism, kinds of clinical values in medical thinking, and value certainty. Chapter 5 asks the question "How is breast cancer a social phenomenon?" Here the focus is on how our understanding of breast cancer entails social and political dimensions. Specifically, the chapter explores the philosophical problems of collaboration, medicalization, and politicalization. Chapter 6 asks the question "What is the relation among the descriptive, explanatory, evaluative, and social dimensions of breast cancer?" Here the emphasis is on how these dimensions of breast cancer mutually define and situate each other within particular contexts. Given this, breast cancer is seen to be an integrative phenomenon that is best understood and treated through a personalized or precision-based medical approach. Chapters 7 asks the question "What are the ethical implications of understanding breast cancer?" Here the focus is on issues discussed in biomedical ethics, namely, informed consent, risk assessment, and just access to medical care and how these issues arise in response to our understanding of breast cancer. Chapter 8 draws the investigation to a close with lessons learned from this philosophical reconstruction of breast cancer.

As this inquiry shows, philosophy teaches us a number of lessons about how we understand and treat breast cancer as well as how we may wish to change our views and practices in breast cancer medicine in order to improve breast cancer care. Alternatively, a study of breast cancer teaches us a number of philosophical lessons about our understanding of reality, knowledge, and the values that guide us. This display of the rich integration of philosophy and medicine and the questions they ask remind us that clinical understandings of disease are not found in a vacuum. They are rather set within contexts of knowing and doing that evolve with time. So we begin with a philosophical analysis of breast cancer inspired by my journey as a breast cancer patient and as a philosopher of medicine who wishes to show the relevancy of philosophical thinking in medicine today and to share thoughts about how to navigate the uncertainty of breast cancer diagnosis, prognosis, and treatment.

Chapter 2

How Is Breast Cancer Described?

PERSONAL MUSINGS

As a breast cancer patient, I often asked myself about what was being described to me by my breast cancer specialists. What was this clinical condition that I was said to have that required clinical attention? Given that I was asymptomatic and experienced no signs or symptoms for breast cancer except a lump that was determined for years by radiologists to be benign, was the disease I was said to have "real"? If real, in what way was it real? Was the breast cancer composed of matter, which meant it had a specific position in time and space? If so, what and where was it? Was the cancer in the cells in my right breast, something more systematic within my body, something that originated outside my body, something in my inherited family line, all of these, none of these, or something else? What was this condition called "breast cancer" that expressed itself in a specific way (e.g., ER+, HER2+, BRCA1)? Given that breast cancer comes in different kinds, what is the relation between the general classification of "breast cancer" and its many kinds (e.g., ER+, HER2+, BRCA1)? Further, what is the relation between a kind of breast cancer and *my* breast cancer? To what extent can the general classification of "breast cancer" be reduced to a particular kind of breast cancer and to *my* breast cancer? If reduction is possible, which it seems it is, given the

	realism	antirealism
(a)	realism (the view that reality is independent of human perception and thought)	antirealism (the view that reality depends on human perception and thought)
(b)	physicalism (the view that reality is physical)	idealism (the view that reality is ideas)
(c)	reductionism (the view that complex phenomena are explainable in terms of their parts)	holism (the view that the whole is greater than the sum of its parts)

Figure 2.1 Ontological issues.

way clinicians think about breast cancer, what is entailed in such a reduction? In what way does the part reflect the whole? What is lost in such a reduction? In what way does the part not reflect the whole? Can I ever be sure that the description of my breast cancer is accurate? Why should I care about such questions, especially if the breast cancer does not return? Herein lay the basis of some of my philosophical reflections on how breast cancer is described and the ontological or reality-related issues concerning realism, physicalism, and reductionism. In what follows, and in the context of discussing the nature of breast cancer, realism is contrasted with antirealism, physicalism with idealism, and reductionism with holism. Figure 2.1 offers clarification and reference.

REALISM

Breast cancer is understood, appreciated, and seen through a set of descriptions. Here "description" refers to the process of giving an account of something through language or other representations.

In medicine, description takes place typically by providing "facts." The term "fact" is derived from the Latin root *factum* and refers to "a thing done" or a "reality of existence." It refers to something that has "really" occurred or is "actually" the case. In the case of breast cancer, the American Cancer Society describes breast cancer in terms of the facts of cellular change that occurs in the breast. As it says, "[b]reast cancer starts in abnormal cells in the breast. These cells are very different from normal, healthy cells. They begin to grow out of control and produce more cells that develop into tumors, or growths. These cancer cells can multiply in the breast and spread to other parts of the body" (2011, 4).

To understand breast cancer, then, it is important to understand what is the breast and what is cancer. In females, the breast is "either of the pair of mammary glands extending from the front of the chest in pubescent and adult human females and some other animals" ("Breast" 2014) or "either of the analogous but rudimentary organs of the male chest especially when enlarged" ("Breast" 2014). In females, the breast is composed of fifteen to twenty sections or lobes. Each lobe is made of many lobules, tiny glands that can make milk. When a female produces milk, milk flows from the lobules through ducts or thin tubes to the nipple. Fibrous tissue (i.e., that which supports and connects tissue and organs) and fat fill the spaces between lobules and ducts (National Cancer Institute 2014d, 1). In males, breast cancer "is fundamentally identical to breast cancer in women with a few exceptions" (Selleck and Tiersten 2004, 646). For one, males have much less developed breasts than females. For another, males tend to have less fat tissue than females (although this is changing in developing countries with the increase of obesity among males). Otherwise, breast cancer in females and males has remarkably similar clinical presentation, pathologic appearance, response to treatment, and overall prognosis (Selleck and Tiersten 2004, 655).

According to the National Cancer Institute, cancer is "[a] term for diseases in which abnormal cells divide without control" (National Cancer Institute 2014d, 36). "[C]ancer begins in cells, the building blocks that make up all the tissues and organs of the body, including the breast" (National Cancer Institute 2014d, 2). Normal cells grow and divide to form new cells as needed. Old or damaged cells die and new cells take their place. When this process errs, new cells form when they are not needed and old or damaged cells do not die as they do in a normal process. The buildup of extra cells can form a mass called a "tumor," "growth," or "lump" (National Cancer Institute 2014d, 2). On this view, cancer is uncontrolled cellular division and survival. In this sense, cancer can be seen to be an "immortal" state of the body, as they continue to replicate without limits (Skloot 2010). Although this may not represent what we usually mean by "immortal," the idea is that what marks cancer cells is their ability to survive and grow in what appears to be endless ways.

It is important to recognize that not all tumors are the same. Those that are not harmful to life and do not spread are called "benign." Those that are expected to develop into cancer and cause problems are called "suspicious" or "precancerous." Those found in the lining of tissue but have not spread are called "carcinoma in situ," where "carcinoma" comes from the Greek root *karkinos,* meaning "crab," and the Latin root "in situ" meaning "in its place." Those that are cancerous are called "malignant" and can be harmful to life, spread to nearby and distant organs and tissues, and, even if removed, can grow back. In the case of breast cancer, after spreading through the lymph system and blood vessels, cells may attach to other tissues and grow to form new tumors that may damage other tissues (National Cancer Institute 2014d, 2–3). Other tissues can include nearby lymph nodes and those above the collarbone, in the chest behind the breastbone, and in the lung or bone (National Cancer Institute 2014d, 10). These tissues, even though they may be located outside the breast, still involve the same biological kind of abnormal cells as found in the primary or original tumor. This type of spread is called "metastatic breast cancer" (National Cancer Institute 2014d, 3), where "metastatic" (from the Latin roots *meta*, meaning "beyond," and *static*, meaning "stillness") refers to the spread of cancer from one part of the body (e.g., a breast) to another (e.g., bones).

On this view, breast cancer is "real" in the sense that it is an actual object independent of the clinician observing it. In philosophy, this view about reality is known as *metaphysical realism*. A metaphysical realist holds that observable objects and events are "actual" and independent of the person observing them. While there are numerous versions of metaphysical realism, a metaphysical realist typically claims that (1) that which is described by a subject matter (S) exists, (2) its existence is independent of the knower observing it, (3) claims made about S are "true" about the world, and (4) "truth" is framework independent. Realism is associated with the thinking of Ancient Greek philosopher Plato (428–348 B.C.E.), who holds that reality is composed of unchanging essence. As Plato says in the *Timaeus*, "What is that which always is and has no becoming . . .? That which is apprehended by intelligence and reason is always in the same state" (Plato 1961, 27D–28A). For Plato, reality is "always is," stable and consistent (i.e., in the "same state"), and out there to be discovered by "intelligence and reason." Of course, for Plato, there are aspects of reality that change (e.g., with age or injury), but the essence of reality (e.g., humanness) does not change.

In the case of breast cancer, it is safe to say that most clinicians and patients hold that breast cancer "really" exists and its existence is independent of the conceptual structures and linguistic formulations that clinicians and patients use to describe it. Breast cancer is breast cancer and not lung cancer, although there can be variations on what kind of breast cancer it is. It is also safe to say that most clinicians and patients hold that the statements that are made about breast cancer are true and that the "truth" of breast cancer is framework independent. They are true insofar as they are based on clinical evidence grounded in objective or unbiased criteria used to evaluate clinical knowledge claims about breast cancer.

Philosophers of medicine call this realist view of disease "ontological" (Engelhardt 1981 [1975], 33). On an ontological view of disease, disease is understood as a "substantial thing" (Cutter 2003, 32) that is regarded as an entity in itself distinguishable by specific changes and causes. As Engelhardt says, "The *ens*, the being of the disease, can be variously understood as either a thing, or a logical type, or both. Medical ontology in the strong sense refers to views in which disease is conceived as a thing, a parasite, in contrast with 'Platonic' views of disease entities in which diseases are understood as unchanging conceptual structures" (Engelhardt 1981 [1975], 33). On this view, disease is an entity that can be said to exist in some sense outside the manifestations of a particular individual (Aronowitz 2007, 13) and expresses itself in similar ways in patients.

The Swiss physician Paracelsus (1493–1541) supports an ontological view of disease (Pagel 1958, 137). As he says, "Diseases are, therefore, entities, each in its own right" (qtd. in Pagel 1958, 138). Each disease has its own manifestation in the body, analogous to how organs have their own manifestation in the body. As Paracelsus says, "The 'localis anatomia' . . . determines the seats of disease in man and the organs that are to suffer together and so cause the symptoms" (qtd. in Pagel 1958, 138). On this view, disease is an objective entity defined by specific changes and causes. In the case of breast cancer, breast cancer is an objective entity located in the breast which brings about the signs and symptoms patients experience. It is defined by specific changes in the breasts and specific causes unique to its kind. Breast cancer is a disease in its own right.

Yet our understanding of breast cancer may not fully meet a "realist" or "ontological" standard of reality. While this may sound crazy to clinicians and patients alike, bear with me for a moment. How we understand breast cancer is dependent upon frames of reference through which we view breast cancer. That is, the phenomenon we call "breast cancer" is in part a construct based on frameworks through which it is described. Tying back to the previous discussion, while breast cancer may be said to "exist," it does not "exist" independent of the clinician, researcher, or patient as knower and doer. Claims made about breast cancer are dependent upon a knower and doer. They reflect how humans know in medicine, and those ways are reflective of prevailing theoretical and practical contexts through which the clinical phenomenon is seen and treated. Those ways are reflective of how humans engage in their world as actors, engineers, and caretakers. In this way, the "truth" about breast cancer is dependent upon the frameworks and content through which clinicians, researchers, and patients understand and manage it.

Consider very generally how our descriptions of breast cancer are dependent upon at least eight frames of reference. First, breast cancer may be described as a *syndrome,* or collection of signs and symptoms. Its signs include a lump or thickening in the breast, skin changes in the texture or color of the breast, an inverted nipple, and/or nipple drainage (National Cancer Institute 2014a). Second, breast cancer may be described in terms of an *etiological* or causal framework. Increasingly, breast cancer is described in terms of what brings it about, such as "large amounts of a protein called HER2" (National Cancer Institute 2014d, 4). Third, breast cancer may be described in terms of a *treatment response.* The stages and grades of breast cancer in the current system of clinical classification are arranged in order to aid the "prognosis of patients with newly diagnosed breast cancer" and "to prevent futile therapy" (American Joint Committee on Cancer 2010, 422–423). Fourth, breast cancer may be described in terms of *linkage to genetic mutations.* Tumors with certain kinds of genetic mutations (e.g., BRCA1) are distinguished from tumors with other kinds of genetic mutations (e.g., BRCA2) (National Cancer Institute 2015c). Fifth, breast cancer may be described in terms of an *immunological condition.* Here, breast cancer is a consequence of a patient's innate and adaptive immune response and the immune escape mechanisms employed by tumor cells (Curigliano et al. 2007). Sixth, breast cancer may be described in terms of *linkage to interacting environmental factors.* Evidence indicates that lifestyle (e.g., diet) and environmental (e.g., carcinogenic) factors contribute to estrogenic activity and to the incidence of breast cancer (National Institute of Environmental Health 2012, 2). Seventh, breast cancer may be described in terms of a *gender-specific* (or as some prefer, *sex-specific*) *condition* (Selleck and Tiersten 2004, 648). As previously mentioned, less than 1% of breast cancer cases are found in males, making breast cancer a gender-specific disease that primarily affects females. Eighth, breast cancer may be described in terms of an *ethnic-specific condition.* As previously mentioned, the incidence rate for breast cancer for whites is higher than for blacks, Asians/Pacific Islanders, American Indians/Alaska Natives, Hispanics, and non-Hispanics, and yet blacks have a higher mortality rate from breast cancer than whites (National Cancer Institute 2015f, 1–2). In short, breast cancer is described in terms of various descriptive frameworks, including syndromatic, etiological, therapeutic, genetic, immunological, environmental, gender-specific, and ethnic-specific.

In some sense, then, breast cancer is a function of the descriptive frameworks through which we understand a phenomenon. In philosophy, this view about reality is known as *metaphysical antirealism.* There are various versions of the view that descriptions are dependent on contextual frameworks. For instance, (1) the *skeptic* (e.g., Greek philosopher Sextus Empiricus [approximately 160–210 C.E.]) denies that that which is described by a subject matter (S) exists. The (2) *idealist* (e.g., German philosopher Gottfried Wilhelm Leibniz [1646–1716]) holds that reality is fundamentally mental in nature and denies that its existence is independent of the knower observing it. The (3) *instrumentalist* (e.g., German philosopher Andreas Osiander [1498–1552]) affirms that a theory is regarded as an instrument for producing predictions or techniques for controlling events and denies that the claims made about S are "true" about the world. The (4) *constructivist* (e.g., Dutch-American philosopher Bas van Fraassen [1941–present]) holds that a theory gives us at best empirical support for something and denies that "truth" is framework independent. In the case of breast cancer, a skeptic might say that, while particular cases of abnormal cellular growth and survival in the breast exist, breast cancer in general does not exist. An idealist might say that the phenomenon of breast cancer is dependent upon a knower's view of it. An instrumentalist might say that the phenomenon of breast cancer serves as an instrument for producing predictions and interventions for managing uncontrolled cell division and growth of breast cells. A constructivist might say that our notion of breast cancer is nothing more than a conceptual placeholder for the empirical observations we have of uncontrolled breast cell division and growth.

The prior listing of metaphysical antirealist positions from the history of philosophy can be overwhelming and so I'll offer a more streamlined way to look at this view of reality in general, and of breast cancer in particular, drawn from the philosophy of medicine literature. Recall the previous discussion of the "ontological" view of disease. Early philosophers of medicine, such as Rudolf Virchow (1821–1902) (1958 [1895], 192) contrasted an "ontological" view of disease with a "physiological" or functional view of disease. On a "physiological" view, disease is understood in terms of particular departures from general physiological norms (Engelhardt 1981 [1975], 36). It is always expressed as it is found in particular patients as individual departures from accepted physiological norms. On this view, disease is not a thing out there to be discovered. Rather, disease is a relation, "the result of individual constitutions, the laws of pathology, and the peculiarities of environments" (Engelhardt 1981 [1975], 36; see also Reznek 1987, 169). It is a relation among the criteria used to frame and name it. In short, disease is a constructed reality that brings together ways we understand it.

In the case of breast cancer, breast cancer is understood as a relation among uncontrolled cellular growth in a particular patient's breast cells, an understanding regarding what constitutes normal versus pathological cell growth, and observations of environmental forces that contribute to pathological cell growth. On a physiological view of breast cancer, the universal of breast cancer does not exist; the predicates of breast cancer do. That is, the general notion of breast cancer does not exist; particular expressions of breast cancer do. Breast cancer is ER+, HER2+, or BRCA1, or whatever the kinds of breast cancer are determined to be at a particular time. It is Stage 0, I, II, III, or IV, and Grade 1, 2, or 3, or whatever the stages and grades are determined to be at a particular time. (For more on these distinctions, see Chapter 3) Breast cancer is *my* breast cancer, which is to be distinguished from *your* breast cancer. (For more on a personalized medical view, see Chapter 6.) Various kinds of breast cancer express particular processes in the body as well as particular relations between the patient and her environment. Here, the truth of breast cancer is framework dependent and, when the framework changes, so does the "truth" or claim.

This is not to suggest that the general category of "breast cancer" is not "real" in any sense. On a "physiological" view of disease, the general category of "breast cancer" describes shared factors of a clinical condition expressed in the breast, or whatever context is specified (e.g., the genome). It describes shared signs of the breast and shared symptoms experienced by breast cancer patients. It describes the etiological, therapeutic, genetic, immunological, environmental, gender-specific, and ethnic-specific factors that are shared by particular expressions of breast cancer. The general category of breast cancer, or rather the grammar we use to describe breast cancer, as Wittgenstein might say, "tells us what kind of an object anything is" (Wittgenstein 1963, sec. 373). It gives a name or label to "the possibility for the coherence of human meaning" (Engelhardt 1996, 96). While the "coherence" of human meaning may not be the "essence" of what something is, it brings together the sensory observations and interpretations that are identified with what we experience. In this way, a "physiological" view of disease in general, and breast cancer in particular, provides a stable and reliable basis for observation, testing, and intervention in the clinical setting. Another way coherence is brought to descriptions of reality is through accounts of the material components of nature, a topic that is next considered.

PHYSICALISM

A dominant view in medicine today is that breast cancer is a material or physical phenomenon. As seen in the previous section, and according to the American Cancer Society and the National Cancer Institute, breast cancer begins in breast cells. When there are errors in cellular growth and regulation, new cells form when they are not needed and old or damaged cells do not die as they do in a normal process. This is a condition breast cancer specialists call "hyperplasia," that is, "growth by virtue of cells increasing in *number*" (Mukherjee 2010, 15) or "atypia," that is, an irregular condition. When such cell growth occurs outside the tissue, it is called "invasive" or "malignant" and can be harmful to life, and spread to nearby and distant organs and tissues. When growth is expected not to be malignant, it is called "benign," "suspicious," or "precancerous," in the order of increasing probability that the growth is considered cancer.

The cells that make up breast cancer are different from the cells that make up, say, lung cancer. While all cells of the body contain the same genomic content, cells differ in terms of how they differentiate in the body (Thibodeau and Patton 2005, 58–60). In other words, cells differ in terms of what is genetically "turned on" and "turned off," so to speak, thereby leading certain cells to generate different kinds of cells that function in specific ways in the body. As an example, breast cells can be ductal, lobular, or medullary; lung cells can be goblet, ciliated, or basal cells. In this way, breast cancer cells differ from lung cancer cells, although one can have breast cancer in the lung, a condition called "metastatic breast cancer" (National Cancer Institute 2014d, 3).

Up through the mid-twentieth century, breast cancer was thought to be a single clinical condition (Van Epps 2012, 44). Today, this is not the accepted view, and there are a number of ways to describe the physical nature of breast cancer. A common way to describe breast cancer is in terms of *anatomic and pathological standards* using criteria that apply to the anatomy and tissue structure of the breast. The most common type of breast cancer is "ductal carcinoma." Ductal carcinoma begins in cells that line a breast or milk duct. "About 7 of every 10 women with breast cancer have ductal carcinoma" (National Cancer Institute 2014d, 3). The second most common type of breast cancer is "lobular carcinoma." Lobular carcinoma begins in the lobules or milk-secreting gland and "1 out of every 10 women with breast cancer has lobular carcinoma" (National Cancer Institute 2014d, 3). If the carcinoma or irregular cell growth stays within the ducts or the lobules, the carcinoma is considered "noninvasive" and is labeled either "ductal carcinoma in situ" (DCIS) or "lobular carcinoma in situ" (LCIS), although there is a debate about whether either of these conditions should be called "carcinoma" or be part of the "cancer" staging system (for more on this, see Chapter 6). If the carcinoma spreads beyond the ducts or lobules, it is considered "invasive" and called "invasive ductal carcinoma" (IDC) or "invasive lobular carcinoma" (ILC), respectively. Such a condition is considered "cancer."

If the carcinoma spreads beyond the breast into other areas of the body, it is called "metastatic breast carcinoma" or "metastastic breast cancer."

Other types of breast cancer that are less common include medullary (where the tissue is a soft, fleshy mass resembling the medulla of the brain), mucinous (where the cells float in pools of mucin, the key ingredient in mucus), papillary (where the cells have small, finger-like projections), tubular (where the cells are made up of tube-shaped structures called tubules), cribriform (where the cells invade the breast stroma or connective tissue), and inflammatory (where the cells invade the breast lymph vessels and lead to swelling in the breast) ("Types of Breast Cancer" 2013). As seen here, a common way to describe breast cancer is in terms of the kind of tissue it is composed.

Another common way to describe breast cancer is in terms of *molecular types* called biomarkers or biological markers. Biomarkers "are proteins that are either produced by cancer cells themselves or by other cells in response to cancer" (Patient Resource 2011, 15). A biomarker is a broad term that refers to deoxyribonucleic acid (DNA), ribonucleaic acid (RNA), protein, metabolite, or lipid material. A biomarker is a measurable substance in an organism whose presence is indicative of some phenomenon, such as disease, infection, or environmental exposure (Gardell 2014). "Most protein biomarkers are used to monitor response and/or detect occurrence or progression during follow-up after treatment. Some protein biomarkers are used to help guide treatment decisions" (Patient Resource 2011, 15). Among the most well-known biomarkers that guide treatment for breast cancer are estrogen receptors, progesterone receptors, and human epithelial growth factor receptor 2 (HER2).

In 1968, Chicago chemist Elwood Jensen found the estrogen receptor (ER), the molecule responsible for binding estrogen and relaying its signal to the cell (Mukherjee 2010, 215). In the 1970s, a number of researchers identified the progesterone receptor (PR), the molecule responsible for binding progesterone and relaying its signal to the cell (Brinton et al. 2008, 314). High levels of ER or PR (indicated by ER+ or PR+, respectively) identify tumors most likely to increase in size in the presence of hormones and to respond to hormone therapy. In the case of ER+ breast cancer, which comprises approximately three-fourths of all breast cancer cases, the hormone estrogen binds to an estrogen receptor, enters a cell, and turns on genes that promote uncontrolled cell division and survival. The same process occurs in the case of PR+ breast cancer, but the hormone that binds to a progesterone receptor is progesterone, which enters a cell and turns on genes that promote uncontrolled cell division and survival (Van Epps 2012, 44). Why estrogen and progesterone receptors act this way in cases of breast cancer has yet to be understood.

Another well-known biomarker for breast cancer is human epithelial growth factor receptor 2 (HER2). German scientist Axel Ullrich (1943–present), working at California-based Genetech, isolated HER2 in the 1990s (Mukherjee 2010, 415). UCLA oncologist Dennis Slamon (1948–present) found an association between certain cancer cells and the presence of HER2, thus bridging the gap between an aggressive type of breast cancer and an oncogene (Mukherjee 2010, 416). An oncogene is a very broad and diverse concept. In general, the "normal" gene that encourages cell division is called the proto-oncogene. Changes to the "normal gene" can be manifest in many different ways so that it becomes an oncogene, a mutation that is either disabling or that modifies the function of the gene product (Gardell 2014). In normal cells, HER2 sends signals into the cell that trigger division and survival in response to growth factors. When HER2 levels are elevated, as occurs in 20% to 30% of breast cancer cases, it acts like a switch that is always on (Val Epps 2012, 44). High levels of HER2 (indicated by HER2+) identify tumors most likely to increase in size in the presence of

HER2 protein and to respond to anti-HER2 therapy (e.g., the drug Herceptin [trastuzumab]) and to some specific chemotherapeutic drugs (e.g., Xeloda [capecitabine]). The discovery of HER2 in cancer medicine has promoted more work on the underlying genetic makeup of cancer, including the pathways that are unique to certain kinds of cancers, even those found in different anatomical and physiological areas of the body.

There are other types of breast cancer with distinct biomarkers. Basal or triple-negative breast cancer has cells that are negative for three common biomarkers: ER, PR, and HER2. Triple-negative breast cancer is aggressive and challenging to treat because it does not respond to the therapies developed for ER+, PR+, and HER2+ breast cancer. The gene expression pattern of this cancer is similar to cancer cells found in the deeper basal level of breast ducts and glands (Susan G. Komen 2014b). Other biomarkers, such as cancer antigen (CA) 27-29, CA 15-3, and urokinase-type plasminogen activator (uPA), are currently used in conjunction with diagnostic imaging, patient history, and physical examination to detect breast cancer.

Increasingly, work is being done to describe breast cancer in terms of its *genetic mutations*. By definition, a gene (from the Greek root *genos*, meaning "offspring") is the smallest functional genetic unit. Among other things, a gene encodes RNA, which exerts a biological effect directly or encodes a protein (Gardell 2014). In the 1990s, mutations in the BRCA1 (breast cancer 1) and BRCA2 (breast cancer 2) genes were among the first cancer-related gene mutations to be identified (Patient Resource 2011, 18). Mary-Claire King (1946–present) and Mark Skolnick (1946–present) from the University of Utah are credited with such a discovery in the 1990s (Olson 2002, 254–255). BRCA1 is located on chromosome 17 and BRCA2 on chromosome 13 (Selleck and Tiersten 2004, 650). BRCA1 and BRCA2 are tumor suppression genes and, when rendered ineffective by mutation, can contribute to breast cancer by not inhibiting cell division (Thagard 1999, 32). The public learned more about BRCA1 following actress Angelina Jolie's public testimonies about her choices following her own diagnosis of BRCA1 and the treatment decisions that followed (2013, 2015). Tumors with BRCA1 and BRCA2 mutations have distinct characteristics that may make some targeted therapy more effective than traditional treatment strategies. Tumors with BRCA1 and BRCA2 mutations appear responsible for 5% to 10% of all breast cancer cases (National Cancer Institute 2015a). Risk factors for BRCA1 and BRCA2 include age at time of cancer diagnosis (before the age of 50), family history (including breast and ovarian cancer), identification of gene mutation in family member, and ethnic background (Ashkenazi or Eastern European Jewish heritage) (Patient Resource 2011, 18). Mutations in BRCA1 or BRCA2 account for approximately 15% of ovarian cancers (National Cancer Institutes 2015a).

As one can imagine, the hunt is on for the additional genes that lead to breast cancer. In 2014, Dr. Marc Tischkowitz and colleagues at the University of Cambridge linked a third gene, PALB2, to breast cancer (Bakalar 2014). PALB2 stands for Partner and Localizer of BRCA2. PALB2 is a protein encoded by the PALB2 gene, which binds to BRCA2 and increases the chance of developing breast cancer. Another focus of research lately is p53, also called the TP53 gene. The gene p53 provides instructions to the body for making a protein that stops tumor growth. An understanding of the genes that are contained in chromosomes that are found in cells and that build the protein that controls the structure and function of cells is a major source of attention in breast cancer medicine today.

In addition, the hunt is on to understand the immunology of breast cancer. The focus here is on an aspect of the immune system that allows breast cancer cells to go unregulated by the body (Curigliano et al. 2007). Breast cancer does not result from a foreign agent that invades the body and leads to disease. Breast cancer arises from the body's cells themselves. Because of this, the body's immune system will typically not react defensively against the developing disease condition until much too late. Oncologist Giuseppe Curigliano and his colleagues put the challenge to understand the immunology of breast cancer in this way: "A better understanding of the relation between innate and adaptive immune responses, of the immune escape mechanisms employed by tumor cells, and acknowledgement of the importance of both cell-mediated and humoral adaptive immunity for the control of tumor growth are necessary for leading to a more comprehensive immunotherapeutic approach in breast cancer" (Curigliano et al. 2007, 1).

In the foregoing, and as one would expect today in medicine, breast cancer is described in terms of its material or physical properties. One would expect this because medicine is a science and science studies the physical world, which can be described using empirical methods. In philosophy, the view that reality is physical is called *physicalism* (or *materialism*). While there are various versions of physicalism, a physicalist typically holds that (1) the composition of reality is physical matter, (2) matter is real, (3) matter has a position in space and time, and (4) matter can be quantified. Physicalism is associated with the thinking of British philosopher Thomas Hobbes (1588–1679). As Hobbes says, "The World . . . is Corporeal, that is to say, Body; and hath the dimensions of Magnitude, namely, Length, Breadth, and Depth: also every part of the Body, is likewise Body, and hath the like dimensions" (Hobbes 1950, ch. 46). Viewing the world as corporeal, or as physical matter, provides modern science a tangible way to quantify, measure, and know reality, and thereby offer a stable foundation for a scientific and medical understanding of reality.

In the case of breast cancer, it is safe to say that breast cancer specialists hold that breast cancer is composed of physical matter, and specifically mutated cells within certain anatomical and physiological regions of the body. Such matter can be observed in time and space with diagnostic technologies and can be quantified according to clinical and pathological standards developed by the American Joint Committee on Cancer (AJCC) (Patient Resource 2011, 20). Here *clinical* standards are based on characteristics of the tumor assessed by physical examination, x-rays, and/or imaging studies and laboratory results. *Pathological* standards are based on a pathologist's examination of tissue specimen removed during a biopsy or surgery. (For more on the standards of the AJCC, see Chapter 3.)

Mechanistic physicalism or materialism has a long tradition in philosophy and medicine. According to this view, objects in the world are composed entirely of matter and motion and located in space and time. They organize and operate according to a finite number of fixed physical laws, or as some say, the laws of a machine. For mechanistic physicalists, the human is a machine. As eighteenth-century French physician Julien Offray de La Mettrie (1709–1751) puts it in *Man and Machine*, "Let us then conclude boldly that man is a machine, and that in the whole universe there is but a single substance differently modified" (1912, 141). Nineteenth- and twentieth-century oncologists, such as Joseph Lister (1827–1912) and Dennis Slamon (1948–present), respectively, approached their work as if the human is a machine and provided stunning breakthroughs in our understanding of cancer. Lister's finding that a mastectomy delayed death from breast cancer (Olson 2002, 46) and Slamon's discovery that high HER2 levels correlate with breast cancer (Mukherjee 2010, 416) came about through viewing the body as a machine and engaging in rigorous empirical testing of treatments on breast cancer patients. Twenty-first-century medical thinking continues to support viewing the human as a machine, disease as the fault in the machinery, and treatment as the tool to fix the machine (Gøtzsche 2007, 42).

Yet our understanding of breast cancer may not fully meet the strict standards of a physicalist or materialist view of reality. Although this may sound quite strange to clinicians and patients alike, there is a sense in which breast cancer is nonphysical. As discussed in the last section, breast cancer specialists and their patients bring to their understanding of breast cancer conceptual frameworks that are not explicitly evidenced in the material world. Take, for instance, an etiological or causal account of breast cancer. As an etiological concept, breast cancer involves not simply a material cause and effect, as in the case of the mutation in the BRCA1 gene and the resulting breast cancer, but the *idea* that something brings something about. While the cause and effect of breast cancer can be described using physical language, the *idea* of the connection between the cause and the effect is not a quantifiable concept. It is a qualitative notion in the mind about the contiguous, sequential, and necessary relation between two phenomena. As Scottish philosopher David Hume (1711–1776) says in *Enquiry Concerning Human Understanding* about the inability to find empirical evidence for the existence of cause and effect:

> We have sought in vain for an idea of power or necessary connexion in all the sources from which we could suppose it to be derived. It appears that, in single instances of the operation of bodies, we never can, by our utmost scrutiny, discover anything but one event following another, without being able to comprehend any force or power by which the cause operates, or any connexion between it and its supposed effect. (Hume 1894, sec. 7, pt. 1)

The point here is that the notion of causal relation is not an empirical concept. We can find at best evidence for a temporal sequence of empirically verifiable objects between what brings something about and the resultant consequence. Applied to breast cancer, breast cancer is an idea that results from a connection between temporally sequential events that are empirically verifiable. (For more on breast cancer causality, see Chapter 3.)

In philosophy, the view that reality is an idea is called *idealism* (or, as some prefer, *idea-ism*). Idealism does not refer to a utopian or optimist view about the world. It is a metaphysical theory about the ultimate nature of reality. It is the view that objects are constituted by the mind and its ideas. An idealist position is expressed by Irish philosopher George Berkeley (1685–1753), who holds that "*esse is percipi*," that is, "to be is to be perceived" (1954, Principles #3). As he says in his *Treatise Concerning the Principles of Human Knowledge*, "things perceived are immediately perceived; and things immediately perceived are ideas; and ideas cannot exist without the mind; their existence therefore consists in being perceived" (1954, 76). For Berkeley, everything is dependent for its existence upon perception, and apart from the perceiving knower, nothing exists. In the form of a syllogism or argument, the view is that (1) ideas exist only in the mind, (2) all things are ideas, and therefore, (3) all things exist only in mind. Of course, a positive diagnosis of breast cancer is to be taken seriously and actions are warranted to treat it.

At first, it may appear that idealism would have little support in medicine and philosophy of medicine. It seems far-fetched and downright irresponsible to say that breast cancer is only in the mind. The last thing a philosopher needs to do is to turn back the hands of time to a period in medicine when many of the diseases and illnesses brought to the clinic by women were considered "in their own heads" (Tuana 2006). The last thing I would want to do as a breast cancer patient is to advocate for views that are wrong or irresponsible and practices that are harmful to patients in breast cancer medicine. Of course, breast cancer is not "in one's mind."

Yet, if one takes a closer look, one can find a version of idealism in the clinical understanding of breast cancer. In philosophy of medicine, this version is known as *nominalism*. Nominalism is the view that that which exists exists in the particular and is given a name that reflects characteristics that are shared with other particulars. As philosophers of medicine Henrik Wulff, Steg Andus Petersen, and Raben Rosenberg say, "[t]here are no genera and species of diseases, and disease names may be regarded as labels which we attach to groups of patients which resemble each other in those respects which we consider important" (1986, 77). On this view, disease is regarded as a name (*nomen*), which is attached to a bundle of particular expressions of a condition that is seen to be important because it is worthy of being removed, prevented, or changed. There are no general diseases out there in reality; there are only particular expressions of biological dysfunction and deformity experienced by patients. As historian of medicine Charles E. Rosenberg (1936–present) says, "disease does not exist until we have agreed that it does, by perceiving, naming, and responding to it" (Rosenberg 1992, xiii). There are no diseases; there are only sick patients.

In the case of breast cancer, a nominalist holds that breast cancer is a name we give to a problem seen in the clinic, named in the clinic or research lab, and deemed to be an appropriate target of medical intervention. It is expressed in individual patients and has characteristics that are shared with cases found in other patients. Yet, in the end, each case of breast cancer is unique and the names we give to different types of breast cancer (e.g., ER+, HER2+, BRCA1) represent labels that cluster various expressions of breast cancer together. As physician and geneticist Bert Vogelstein (1949–present) says, "[e]very patient's cancer is unique because every genome is unique. Physiological heterogeneity is genetic heterogeneity" (Vogelstein, as quoted in Mukherjee 2010, 452). While Vogelstein is far from being a nominalist, his comment emphasizes the lack of the availability of a singular empirical account of cancer. It accounts for the different ways patients with the so-called same cancer have different cancers.

We are faced, then, with the question about what holds individual physical expressions of breast cancer together. As stated in the previous section, the term "breast cancer" gives a name or label to that which brings together various dimensions of human experience, observations, and meaning. It correlates constellations of signs and symptoms for purposes of explanation, prediction, and control (Engelhardt 1981, 37). On this view, breast cancer is a *family of diseases* brought together to explain, predict, and control a certain cluster of phenomena of the breast. This view is in keeping with a view held by nineteenth-century American philosopher C. S. Pierce (1839–1914), who says this about our intellectual conceptions: "In order to ascertain the meaning of an intellectual conception, one should consider what practical consequences might conceivably result by necessity from the truth of that conception; and the sum of the consequences of the entire meaning of the conception" (1865, 5.9). Applied to breast cancer, breast cancer is an idea based on observations of the physical world and conceived for purposes of certain practical actions, namely, diagnosis, prognosis, and treatment. It is a concept developed in medicine and sustained by medicine in theory and practice. As theory, observations, and practice change in light of new evidence and experience, the clinical concept changes to adapt to new ways of understanding the concept and managing the phenomenon. Such new ways are typically more specific than previous ways. This move to reductionism is considered next.

REDUCTIONISM

As one would expect in medicine, descriptions of breast cancer have become increasingly more specific. One would expect this because medicine as a science investigates natural phenomena on an increasingly more specific level made possible by new and improved technology that allows observation of phenomena to take place at increasingly minute levels. Consider a very brief history of our understanding of breast cancer. Over time, breast cancer has been understood in terms of (1) an imbalance of bodily humors, (2) a dysfunction in the anatomical organ of the breast, (3) a pathophysiological occurrence in the breast, and, more recently, (4) a biomolecular and genetic dysfunction of the breast cells. Greek physician Hippocrates (fifth century B.C.E.) held that breast cancer erupted from an excess of black bile in the breast (Olson 2002, 12; Mukherjee 2010, 48). Italian physician Giovani Battista Morgagni (1682–1771) thought that breast cancer resulted from curdled breast milk (Olson 2002, 33). German physician Friedrich Hoffmann (1660–1742) proposed that breast cancer started in a lymphatic blockage in the breast that was exacerbated by heightened sexual activity (Olson 2002, 33). French surgeon Jean Louis Petit (1674–1750) saw the root of breast cancer in enlarged lymphatic glands (Olson 2002, 46). Maryland surgeon William Stewart Halsted (1852–1922) held that breast cancer was an abnormality of cells in the breast and therefore recommended the removal of the whole breast and hormone therapy (Olson 2002, 62). Cleveland clinic surgeon George Crile (1864–1943) thought that breast cancer was localized in the blood and called for pharmaceutical approaches to the treatment of breast cancer (Olson 2002 107). National Institutes of Health physician Roy Hertz (1909–2002) saw breast cancer as a consequence of increased use of estrogen in birth control (Olson 2002, 178). University of Utah researchers Mary-Claire King and Mark Skolnick established in the 1990s that breast cancer is a genetic condition brought about by mutations on chromosomes 17 and 13 (Olson 2002, 254–255).

The move from humoral to anatomic to pathological to biomolecular descriptions of breast cancer highlights medicine's move to reduce clinical phenomena to increasingly particular kinds or types. Humoral accounts are replaced by anatomical accounts that are replaced by pathophysiological ones that are currently being replaced by biomolecular, genetic, and immunological ones. Correspondingly, humoral treatments (e.g., those targeted at the balance of black bile) are replaced by surgical interventions (e.g., a mastectomy) that are being replaced with chemotherapeutic agents (e.g., Taxol [paclitaxel]) that are being replaced with targeted therapies (e.g., Herceptin) that will be replaced with immunological and genetic therapies.

In philosophy, the position that objects are to be reduced to their component parts is known as *reductionism*. Reductionism is the view that objects or phenomena at one level of description are explicable in terms of objects or phenomena at a more specific level of description. A reductionist view can entail a number of strategies, such as (1) one theory reduces to another (a position called "theoretical reductionism"), (2) descriptions and explanations reduce to the smallest possible entities (a position called "epistemic reductionism"), and (3) the "whole" reduces to the sum of its "parts" (a position called "ontological reductionism"). Reductionism is associated with the thinking of British evolutionary biologist Richard Dawkins (1941–present) who defends a gene-centered view of natural selection—and of ourselves. As Dawkins says in *The Selfish Gene*, "I shall argue that the fundamental unit of selection, and therefore of self-interest, is not the species, not the group, nor even, strictly, the individual. It is the gene, the unit of heredity" (1976, 12). For Dawkins, accounts of evolution at levels "above" the gene are not scientifically defensible because they do not minimize the influence of speculative thought in scientific inquiry and conclusions.

Yale social scientist and physician Nicholas Christakis (1962–present) addresses the significance of the reductionist move in science: "for the last few centuries, the Cartesian project in science has been to break down matter into smaller bits, in the pursuit of understanding" (2011). Here the Cartesian project is associated with

French philosopher René Descartes (1596–1650) and his view that "man is a machine." In the case of breast cancer, a reductionist today takes breast cancer to be reducible to mutated cells that survive and replicate in an uncontrollable way. The strategy can include a number of approaches: (1) one theory of breast cancer reduces to another, (2) descriptions of breast cancer reduce to the smallest possible descriptions, and (3) "breast cancer" reduces to the sum of its parts. That is, (1) an anatomical and pathological view of breast cancer is increasingly being reduced to a biochemical view. Regarding (2), descriptions of breast cancer target the smallest possible descriptions (e.g., the level of the gene). Regarding (3), "breast cancer" is typically described using anatomical, pathological, and biochemical factors (e.g., Stage 2A, Grade 2, and ER+ breast cancer, respectively). On this view, an understanding of breast cancer takes place at increasingly more specific levels of description and more specific treatments are developed accordingly.

Reductionism garners much support in contemporary medicine. The move from descriptions of signs and symptoms to underlying pathophysiological and pathobiomolecular processes has led to a revolution in medicine in the nineteenth and twentieth centuries and the development of more successful ways to diagnose and treat patients (Foucault 1973 [1963]). As Columbia University oncologist and scientist Sidhartha Mukherjee (1960–present) puts it, reductionism is "supported by the pathologists of the 1940s and 1950s who used simple models of the cell to understand complex phenomena" (2010, 20). Reductionism has gained popularity in medicine because it works. Oncologists replace general clinical descriptions and explanations of breast cancer with more specific descriptions and explanations. For instance, breast cancer is no longer just a tumor in the breast but ER+, PR+, HER2+, BRCA1, or BRCA2 breast cancer—with the recognition that the categories will continue to change with new knowledge. Therapies are increasingly replaced with more specific targeted interventions, again with the recognition that therapies will continue to evolve. For instance, mastectomies are replaced with lumpectomies and targeted pharmaceutical treatments (e.g., Herceptin), with hopes for increasingly more specific and less harmful interventions. A "one-size-fits-all" diagnosis and treatment for breast cancer becomes a thing of the past as specific diagnoses and treatments are developed for specific kinds of breast cancer.

Yet a reductionist account of breast cancer may not accurately reflect the clinical condition we call breast cancer. Reducing breast cancer to a genetic structure misses some important features of breast cancer. To begin with, a mutated gene within a cell is not breast cancer. Breast cancer is composed of mutations within a cell in certain regions of tissue found in an organ called the breast. Such cells operate within immunological and physiological systems of a body that are highly complex. Such a body has the capacities to process information, self-reflect, and feel pain and suffering. Such a biopsychological being lives in a social environment that can contribute to mutations in the cell through, for instance, toxins that affect the being's immunological system. In the end, humans are biopsychosocial beings and not simply biological beings. Reducing breast cancer to abnormal cellular or genetic structure and function limits our understanding of the complex physiological and biopsychosocial events that come together to create what we call breast cancer.

An example that illustrates that the complexity of breast cancer is drawn from current research on breast cancer. In addition to the controls on proliferation brought about by coordinated action of proto-oncogenes and tumor suppression genes, breast cells have at least three other mechanisms that prevent against unconstrained cell division. First, the human body has a DNA repair system that detects and corrects errors in DNA. Should the system fail, the error (now a mutation) becomes a permanent feature in the cell and in all of its descendants. Second, the body's defense system has a way of prompting cells to undergo apoptopsis, or cell death, should some essential component of the cell become damaged or its control system become deregulated. Third, the body's defense system limits the number of times a cell can divide, and so assures that cells cannot reproduce endlessly. The system is governed by a counting mechanism that involves the DNA segments at the ends of chromosome, called telomeres. Telomeres shorten each time a chromosome replicates. Should the telomere not shorten, endless cell proliferation, and thus cancer, may occur. Given this research, breast cancer cells operate within mechanisms of the body (e.g., DNA repair, apotosis, cell replication) that operate along with other mechanisms of the body (Mukherjee 2010) in a living being.

The position in which properties of the whole are not reducible to properties of its individual components is called *holism*. Advocates of holism reject reductionism and hold that the material constitution of an object cannot be explicated in terms of individual material constituents. They deny the necessity of a division between the function of separate parts and the workings of the "whole." Observation cannot be tested in isolation because a test of one observation depends on others, and so on. Holism is associated with the thinking of the Greek philosopher Aristotle (384–322 B.C.E.). As Aristotle says in the *Metaphysics*, "In the case of all things which have several parts and in which the totality is not, as it were, a mere heap, but the whole is something besides the parts" (1941, Bk. H, VIII, 1045a8-10, 818). In other words, as the passage is often reworded, "a whole is more than the sum of its parts."

One can find support for holism in philosophy of medicine. On a holistic view, disease is the abnormal equilibrium of a self-regulating system, as opposed to the end result of a causal chain. As physician and philosopher of medicine Peter Gøtzsche (2007, 42) puts it, disease is composed of physiological mechanisms characterized by complex feedback mechanisms aimed at maintaining normal balance or homeostasis, but unable for a variety of reasons to obtain this goal. In the case of breast cancer, breast cancer is the abnormal equilibrium of cell division and survival in a self-regulating system. While this view could be seen as an expression of a mechanistic view of disease, it parts with it. An understanding of a systematic interaction entails more than simple cause-effect relations; it involves the notion of a *system*, where a system is understood as a set of processes that work together to sustain an event or phenomenon (Marcum 2009). Breast cancer is a process involving mutating breast cells unable to regulate themselves within a human body, thereby permitting uncontrolled replication and survival and operating within the context of interacting bodily processes in the life of a patient who lives in a particular environment.

Similarly, while philosopher of medicine Henrik Wulff and his colleagues view "the mechanical model as an indispensable *part* of the disease concept" (1986, 59), this is not the entire story. As they say, "diseases are not only biological entities. It is not biological organisms, but human beings, who are ill, and even diseases like . . . cancer, which clearly involve biological defect, have causes, manifestations, and effects which reach far beyond the limits of biology" (1986, 59). Disease involves a patient's experience of pain, suffering, and constraints on life goals, and this experience is not an isolated "thing" in the body. The experience of the patient is not incidental to our understanding of disease; it is central. It is central because disease would not be a focus of attention without some level of patient complaints by the patient herself or by a patient in the past. Given this, "the reduction of non-biological phenomena to biology is futile, and, at worst, it leads to a distorted and unacceptable view of man [i.e., a human being]" (Wulff et al. 1986, 59) and the practice of medicine. (For more on the evaluative dimension of breast cancer, see Chapter 4.)

As seen here, "holism" in medicine has come to be associated with approaches that integrate biological considerations with psychological and social ones. The name that is given for this in philosophy of medicine is a *biopsychosocial model* of disease. Psychiatrist George Engel (1913–1999) has been a proponent of this model in medicine. As he says, "a medical model must take into account the patient, the social context in which he lives, and the complementary system devised by society to deal with the disruptive effects of illness, that is, the physician role and the health care system" (1981 [1979], 598). On this view, disease is not just biochemical dysfunction in the body. It entails symptomatic reports by patients, the desires and goals of patients and health care providers, and the social expectations and influences of culture. On this view, breast cancer is not just an anatomical, physiological, cellular, or genetic condition, but one expressed in the life of a patient who lives in a particular environment. On this view, breast cancer is not just a condition of the body, but a condition of a patient within a social context. (For more on the social dimension of breast cancer, see Chapter 5.)

CLOSING

The question "What is breast cancer?" raises a host of descriptive issues in philosophy, including ones about the nature, make-up, and reducibility of reality. What we find is that, initially, breast cancer appears to fit the description of a clinical entity that is real, composed of physical matter, and reducible to its parts. But things are not as simple as one would initially think. Upon reflection, breast cancer is a physical condition framed within prevailing contexts of reference. It is an idea through which clinical facts are seen and interpreted. It is a systematic condition not simply reducible to the sum of its parts. Breast cancer is "real," but perhaps not real in the way we might initially think.

Chapter 3

How Is Breast Cancer Explained?

PERSONAL MUSINGS

As a breast cancer patient, I often thought about the explanations that were provided by my breast cancer specialists for why I had breast cancer. Given that I have no family history of breast cancer and have lived my life following standard lifestyle recommendations for preventing breast cancer, I asked, why did I "get" breast cancer? It seems to me that the answer to the question "why?" depends upon an explanation that appeals to a causal account of my breast cancer's occurrence. If clinicians could understand what caused my breast cancer, they could understand why I "got" breast cancer and they could tailor treatments specifically to it and understand how to prevent a future occurrence. But given that my clinicians could at best provide a limited causal account of my breast cancer, I began to grapple with the uncertainty of my breast cancer diagnosis, prognosis, and treatment. Numerous questions arose. How can I be sure about my own breast cancer diagnosis? What am I to believe about my own prognosis following treatment? Did I really need to have a bilateral mastectomy with lymph removal, interventions that even one of my doctors called "harsh"? Did

I really need to have interventions that carry side effects (e.g., lots of scars, loss of mobility, numbness, and lymphedema) that I will live with the rest of my life? Do I really need to be on adjuvant therapy, "therapy" that leads to its own set of side effects (e.g., stroke, blood clots, bone loss)? Given that my clinicians could at best provide a limited causal account of my breast cancer, to what extent can I be confident about my breast cancer diagnosis, prognosis, and treatment? Alternatively, to what extent can I be skeptical about my breast cancer diagnosis, prognosis, and treatment plan? Obviously, there can be danger here with this type of skepticism about clinical knowledge and practice, especially if one is really diagnosed with breast cancer as I was and one has competent breast cancer specialists as I do. So, to put it another way, what constitutes a healthy sense of skepticism when presented with a diagnosis of breast cancer, its prognosis, and recommendations for treatment? How does a healthy sense of skepticism help navigate the terrain of decisions in breast cancer care? Herein lay the basis of some of my philosophical reflections on how breast cancer is explained and the epistemological or knowledge-based issues of empiricism, causality, and cognitive certainty. In what follows, and in the context of knowing breast cancer, empiricism is contrasted with rationalism, different notions of causal relations are distinguished, an old notion of cognitive certainty is contrasted with a new notion of cognitive certainty, and different notions of cognitive skepticism are considered. Figure 3.1 illustrates the upcoming topics.

While the previous chapter explored the reality of breast cancer, this chapter explores how breast cancer is explained. Readers will find overlaps between the ontological and epistemological issues presented in these two chapters, although there are notable differences in what topics emerge in the different traditions of thought.

(a)

empiricism	rationalism
(the view that knowledge is derived from sense perception)	(the view that knowledge is derived from reason)

(b)

necessary causal relation	sufficient causal relation	contributory causal relation	spurious causal relation
(a factor is required to bring an effect about)	(a factor brings an effect about but another factor could)	(a factor contributes to bringing effect about)	(a factor mistakenly appears to bring effect about)

(c)

old notion of certainty	new notion of certainty
(100% certainty)	(probabilistic certainty)

(d)

absolute skepticism	philosophical skepticism	common-sense skepticism
(doubts possibility of knowledge altogether)	(doubts most cherished claims and assumptions)	(questions claims and assumptions)

Figure 3.1 Epistemological issues.

EMPIRICISM

Breast cancer is understood, appreciated, and seen through a set of explanatory assumptions and claims. Here "explanatory" refers to that which gives reasons for an event or occurrence and answers the question "why?." According to Engelhardt, "[p]roblems are seen as medical problems because they are presumed to be embedded in a pathophysiological, pathoanatomical, or pathopsychological nexus and because the problems are not experienced as removable at the immediate will of the sufferer" (Engelhardt 1996, 208). In this way, problems become medical problems by meeting a host of criteria, including reliability, reproducibility (or precision), and accurateness (Gøtzsche 2007, 17). And so, we begin the analysis of how breast cancer is explained by considering how breast cancer is known.

A dominant way in which breast cancer is explained is through a grading and staging system of classification developed by clinical professional groups. A pathologic *grade* describes how fast or slow the cancer is growing and progressing (Brown and Freeman, with Platt 2007, 73) and is based upon a pathologist's examination of tissue obtained through biopsy. A grading system of classification is used today for all cancers in allopathic medicine. In the early 1900s, German pathologist D. P. von Hanseman (1858–1920) and Boston surgeon Robert P. Greenough began microscopically to classify breast cancer cells by degree of malignancy (Olson 2002, 103). Today, "[t]he histologic grade assigned to a tumor is a way for the pathologist examining the tumor to describe how cancerous cells are arranged in relation to one another and describes some features of individual cells" (Brown and Freeman, with Platt 2007, 73). Such features include the size and shape of the nucleus and the pattern the cells form as they grow and join together. Currently, there are three levels of clinical classification. "A Grade 1 tumor consists of relatively slow-growing cancer cells that look a great deal like normal cells; these are called 'well-differentiated' cells. A Grade 2 tumor is considered 'moderately differentiated' and a Grade 3 tumor is considered 'poorly differentiated.' The higher the grade, the more the cells are 'scattered,' 'abnormal in appearance,' and 'more aggressive in their spread'" (Brown and Freeman, with Platt 2007, 73).

The *stage* of breast cancer describes how far the cancer has progressed (Brown and Freeman, with Platt 2007, 73) in light of the size of the tumor and whether it has spread to lymph nodes or other parts of the body. A staging system of classification is used today for all cancers. In the early 1920s, German surgeons staged tumors according to three levels: Stage I (small tumors), Stage 2 (larger tumors that spread to axillary lymph nodes), and Stage 3 (tumors that spread to the axilla and surrounding tissue) (Olson 2002, 103). Current cancer staging is based on a system advanced by the American Joint Committee on Cancer (AJCC). The system was developed in the late 1950s and last updated in the seventh edition in 2009 (American Joint Committee on Cancer 2010). The eighth edition will be released in 2017 (American Joint Committee on Cancer 2017). Particular recommendations for the staging of breast cancer come from a task force devoted to this type of cancer.

Staging breast cancer begins with determining the tumor (T), node (N), and metastasis (M) (TNM) involvement of the site being studied. The T category provides information on the size and extent of the tumor within the breast. The higher the rating, the larger the size of the tumor. Staging of the N category determines the number of lymph nodes that have evidence of breast cancer cells. A pathologic determination of N indicates how many lymph nodes are involved as well as the amount of tumor cells that are found in the nodes. The closer the lymph nodes are to the breast, the less extensive is the cancer; the farther the lymph nodes are from the breast tissue, the more extensive is the cancer. Staging of the M category indicates the number of sites to which the cancer has spread. It determines whether there is evidence of distant metastasis, or spread of cancer to another part of the body. The most common sites of distant metastases for breast cancer are the bones, lung, liver, and brain (National Cancer Institute 2014d, 10). Staging of the M category can be *clinical* (i.e., determined by clinical examination) or *pathologic* (i.e., determined by microscopic or molecular tests). Either way, the greater the metastases, the poorer the future outcome for a patient.

Once a particular case of breast cancer has been classified according to the TNM staging system, an overall stage is assigned. While the discussion so far has been on how breast cells are described in terms of grades and stages, such descriptions are recast in terms of an overall stage of breast cancer that guides clinical prognosis and recommendations for treatment (American Joint Commission on Cancer 2010, 422–423). Today, the stage of breast cancer is labeled using the Roman numerals 0, I, II, III, and IV, and the Arabic letters A, B, and C. A brief summary of the current stages of breast cancer is provided

*Stage O is called "carcinoma in situ" (National Cancer Institute, "What," 2014, 8). In ductal carcinoma in situ (DCIS), for instance, "abnormal cells are in the lining of a breast duct, but the abnormal cells have not invaded nearby breast tissue or spread outside the duct" (National Cancer Institute, "What," 2014, 8).

*In Stage IA, "[t]he breast tumor is no more than 2 centimeters" (National Cancer Institute, 2014, 8) across and the cancer has not spread to the lymph nodes. Here, 2 centimeters equals about three quarters of an inch, or about the size of a peanut (National Cancer Institute, "What," 2014, 8).

*In Stage IB, "[t]he tumor is no more than 2 centimeters across" and a small number of "[c]ancer cells are found in the lymph nodes" (a condition called "micrometastases" in the lymph nodes) (National Cancer Institute, "What," 2014, 8).

*In Stage IIA, "[t]he tumor is no more than 2 centimeters across, and the cancer has spread to underarm lymph nodes" (National Cancer Institute, "What," 2014, 8) or "the tumor is between 2 and 5 centimeters" (National Cancer Institute, "What," 2014, 9) and "the cancer hasn't spread to underarm lymph nodes" (National Cancer Institute, "What," 2014, 9). As a comparison, five centimeters equals about the size of a small lime (National Cancer Institute, "What," 2014, 8).

*In Stage IIB, "[t]he tumor is between 2 and 5 centimeters across, and the cancer has spread to underarm lymph nodes" or "the tumor is larger than 5 centimeters across, but the cancer hasn't spread to underarm lymph nodes" (National Cancer Institute, "What," 2014, 9).

*In Stage IIIA, "[t]he breast tumor is no more than 5 centimeters across, and the cancer has spread to underarm lymph nodes that are attached to each other or nearby tissue" (National Cancer Institute, "What," 2014, 9), "the cancer may have spread to lymph nodes behind the breastbone" (National Cancer Institute, "What," 2014, 9), "the tumor is no more than 5 centimeters across and "[t]he cancer has spread to underarm lymph nodes that may be attached to each other or nearby tissue" (National Cancer Institute, "What," 2014, 9), or "the cancer may have spread to lymph nodes behind the breastbone but not spread to underarm lymph nodes" (National Cancer Institute, "What," 2014, 9).

*In Stage IIIB, "the breast tumor can be any size, and it has grown into the chest wall or the skin of the breast" (National Cancer Institute, "What," 2014, 9), "[t]he cancer may have spread to underarm lymph nodes" (National Cancer Institute, "What," 2014, 9), or "the cancer may have spread to lymph nodes behind the breastbone" (National Cancer Institute, "What," 2014, 9).

*In Stage IIIC, "[t]he breast cancer can be any size, and it has spread to lymph nodes behind the breastbone and under the arm" (National Cancer Institute, "What," 2014, 10), "the cancer has spread to lymph nodes above or below the collarbone" (National Cancer Institute, "What," 2014, 10), or "the cancer has spread to lymph nodes above or below the collarbone" (National Cancer Institute, "What," 2014, 10).

*In Stage IV, the highest stage of breast cancer, "[t]he tumor can be any size, and cancer cells have spread to other parts of the body, such as the lungs, liver, bones, or brain" (National Cancer Institute, "What," 2014, 10).

Figure 3.2 Stages of breast cancer.

in Figure 3.2 (for more detail, see American Joint Committee on Cancer 2017).

Currently, the TNM staging system provides a standard based on technologies that are widely available and that can be used by clinicians around the world to determine a patient's future outcomes and treatment plan. It is available to clinicians in developed as well as developing countries and permits communication about breast cancer within and across clinical borders. Anatomic staging allows breast cancer clinicians and investigators to link to the past and to study patients who have been diagnosed over the past six decades and more (American Joint Committee on Cancer 2010, 422).

Despite its wide availability and usefulness, the TNM staging system is limited in its ability to forecast the specific onset and severity of a particular kind of breast cancer and to recommend a particular treatment. It is limited because it relies on general categories of the quantity and quality of tissue to forecast future developments. As a consequence, the American Joint Committee on Cancer (2010) recommends that clinicians take into account other factors that relate to the make-up and prognosis of particular kinds of breast cancer. Included here are results of tests for biomarkers, such as ER, PR, and HER2 content (American Joint Committee on Cancer 2010, 423). The main utility of biomarkers tests is found in distinguishing specific kinds of cancers from others and guiding what kind of treatment should be administered.

In addition, multigene expression assays offered by companies, such as Myriad Genetics and Quest Diagnostics (Lee 2013), provide information on a patient's future outcome and course of treatment for particular kinds of breast cancer (Patient Resources 2011, 17; Beil 2014, 28–29). Multigene expression assays rely on data obtained from biomarkers and genes. Positive tests for biomarkers ER, PR, HER2, CA 27-29, and CA 15-3, and genetic tests for BRCA1 and BRCA2 "all play a role in a complex dance involving both prognosis and prediction for the specific therapies" (American Joint Committee on Cancer 2010, 424) for breast cancer. The American Joint Committee on Cancer (2010, 423) forecasts that when tumor biology tests are more available to clinicians and researchers, it is likely that the TNM staging system will undergo significant changes as they incorporate more specific tests for breast cancer. These ways are brought about by tumor biology tests that promise greater diagnostic specificity for breast cancer and provide more personalized and reliable treatment plans for individual patients. (For more on a personalized medical approach to breast cancer, see Chapter 6.)

The prior overview provides the backdrop for discussing assumptions in breast cancer medicine about how medical practicitioners and researchers explain breast cancer. The focus in this section is on a particular debate in philosophy about how knowledge is acquired. On the one hand, there is the view that knowledge is arrived at through sensory experience. This type of view about knowledge is known as *empiricism*. There are varying versions of empiricism, ranging from what I call "weak" to "strong" empiricism. For our purposes, it is sufficient to appreciate that knowledge based on sensory experience assumes that (1) the object of knowledge can be known, (2) it can be known via an empirical method, (3) empirical data are necessary to support the existence of the object, and (4) our understanding of the object changes with new knowledge. Empiricism is associated with the thinking of British philosopher John Locke (1632–1704), who holds that the mind is a "tabula rasa" or "blank slate" and knowledge comes from sense experience (1924, I, 2, 1). As Locke says in *An Essay Concerning Human Understanding*, "When has it all the materials of reason and knowledge? To this I answer, in one word, EXPERIENCE in that all knowledge is founded, and from that it ultimately derives itself" (1924, II, 1, 2). The type of knowledge that derives from sensory experience is referred to in philosophy as *a posteriori* knowledge, or knowledge "after" or based on experience.

In philosophy of medicine, proponents of empiricism abound. Gøtzsche provides numerous examples of how medicine relies on empirical data in the diagnosis of disease. In the case of myocardial infarction, "patients may present various symptoms (e.g., praecordial pain and anxiety), various clinical signs (tachycardia and sometimes a pericardial friction rub), and various paraclinical findings (e.g., certain electrocardiographic changes and changes in the book concentration of different enzymes)" (Gøtzsche 2007, 56, see also

4–5). In addition, "[t]he onset, progression, possible complications and short-term prognosis of the disease" (Gøtzsche 2007, 57) are advanced. Together, this set of assessments is called "the clinical picture" (Gøtzsche 2007, 57) or "clinical examination." It relies heavily on empirical observations, those at the chairside or bedside as well as those made possible by "*sense-extending* instruments" (Wulff et al. 1986, 18), such as an electrocardiograph, magnetic resonance imaging (MRI), and microscope. The use of laboratory microscopes that assess tissue specimens provides what is called pathological evidence that extends the evidence in the clinical examination.

In the case of breast cancer, patients are typically screened for early detection of breast cancer. Screening is a strategy used in a population to identify an unrecognized disease in individuals who do not exhibit signs and symptoms for the condition. Breast cancer screening is designed to identify precursors (e.g., calcification, hyperplasia) or early stages of breast cancer, thus enabling earlier intervention and management in order to reduce mortality and morbidity from the condition. Breast cancer screening consists of a combination of regular clinical breast examination (CBE) and counseling to raise awareness of breast cancer symptoms for women beginning in their 20s and annual mammography beginning at age 40. Recent debates focus on the need for annual screening for women over 40 in part because of the financial costs as well as an increased concern about the overdiagnosis of breast cancer (American Cancer Society 2015a). If a screening comes back positive, the patient is sent for diagnostic testing. Testing involves a targeted method to identify or eliminate those who have or do not have a disease. Breast cancer testing involves a diagnostic mammography as well as other tests such as an ultrasound (or sonography), computed tomography (CT) scan, magnetic resonance imaging (MRI), and positron emission tomography (PET) scans. Medical examination assesses the clinical symptoms (e.g., pain or discomfort in the breast), clinical signs (e.g., a breast lump or bleeding in the nipple), and paraclinical findings (e.g., abnormal results from a mammogram or ultrasound). If initial tests come back positive, a biopsy follows and the status of the initial signs and symptoms is determined by pathological and biochemical tests. If results come back positive for breast cancer, the stage, grade, hormone and perhaps genetic status, and prognosis of the particular kind of breast cancer are determined. This "clinical picture" relies on empirical observations, those at the chairside or bedside, as well as those made possible by "sense-extending instruments," such as a mammogram, ultrasound, and MRI, and those used in the laboratory to assess tissue specimens obtained through biopsy.

It is difficult to argue that empirical data or sense experience matters little in breast cancer medicine. Breast cancer medicine as an applied science relies heavily on that which is observed in order to explain breast cancer. The screening, testing, and diagnosis of breast cancer based on the TNM system of classification relies on empirical observations of the patient. Yet an empirical account of breast cancer may not fully explain breast cancer. Our understanding of disease entails not only empirical knowledge but what philosophers and practitioners of medicine call "rational diagnosis and treatment" (Gøtzsche 2007, see title of book). According to Gøtzsche, disease is known by what is called a "deductive approach" (Gøtzsche 2007, 61). A deductive approach in medicine reasons from "whole" to "part," from pathophysiological knowledge of disease mechanisms to a particular instance of a disease. When a clinician arrives at a preliminary diagnosis of a disease for a patient who has relevant signs and symptoms coupled with a family history, her reasoning is deductive; it is based on a genetic etiological model of the disease. In addition, a disease is known when a clinician recognizes a clinical picture and makes a diagnosis based on "pattern recognition" (Gøtzsche 2007, 62). A pattern recognition composed of signs and symptoms can lead a clinician to make a preliminary diagnosis of a disease, one that is followed by tests to establish the diagnosis or to rule it out. Further, a disease is known when a clinician renders a "probabilistic diagnosis" (Gøtzsche 2007, 62) based on a statistical calculation of what the condition likely will be prior to ordering tests. The signs and symptoms mentioned earlier, coupled with evidence of a positive genetic history for a disease, can lead a clinician to make a probabilistic diagnosis of a disease which then warrants a certain battery of tests to establish the diagnosis. The point is that, although empirical evidence operates in all the examples mentioned, clinicians bring to their diagnosis of a disease rational frames of reference through which they "know" an object or experience.

In philosophy, this type of approach to knowledge is known as *rationalism*. For a rationalist, knowledge of reality is acquired through reason, which is independent of sensory experience. In this way, rationalism provides a foundation for the origin and expression of knowledge beyond the confines of particular observations. Rationalism relies on what is called *a priori* knowledge, knowledge of necessary truths, innate or intuitive knowledge, or the general principles gaining credibility through the use of reason. There are varying kinds of rationalist positions, ranging from what I call "weak" to "strong" rationalism. For our purposes, it is sufficient to appreciate that knowledge based on reason assumes that (1) the object of knowledge can be known, (2) it can be known via a rational method, (3) rational justification is necessary to support the existence of the object, and (4) our understanding of an object remains stable with rational justification. Rationalism is associated with the thinking of French philosopher René Descartes (1596–1650), who held that mathematics provided certain knowledge. For Descartes, the truths of mathematics and the proofs of geometry are certain in that they are untainted by the fluctuations and relativities of sensory experience. As he says, "[m]ost of all was I delighted with Mathematics because of the certainty of its demonstrations and the evidence of its reasoning" (Descartes 1911,

I, 85). And he goes on: "but I did not yet understand its true use, and believing that it was of service only in the mechanical arts, I was astonished that, seeing how firm and solid was its basis, no loftier edifice had been reared thereupon" (Descartes 1911, I, 85). Mathematics provides a method of inquiry that provides certainty.

For rationalists of medical knowledge, the message is that empirical evidence does not operate alone and outside the context of conceptual matrices that frame and organize the data. Consider what Columbia University physician and oncologist Jerome Groopman tells us about how medical thinking is typically taught and how this does not accurately represent how doctors think:

> Medical students are taught that the evaluation of a patient should proceed in a discrete, linear way: you first take the patient's history, and then perform a physical examination, order tests, and analyze the results. Only after the data are compiled should you formulate hypotheses about what might be wrong. These hypotheses should be winnowed by assigning statistical probabilities, based on existing databases, to each symptom, physical abnormality, and laboratory test; then you calculate the likely diagnosis. This is Bayesian analysis, a method of decision-making favored by those who construct algorithms and strictly adhere to evidence-based practice. (Groopman 2007, 11)

Bayesian analysis is named after the eighteenth-century cleric Thomas Bayles (1701–1761) and "takes probabilities to be an estimate of an event's certainty" (Marcum 2008, 106). It "estimates the event's occurrence on prior experience" (Marcum 2008, 106). Baysian analysis expresses the relative probabilities of different states being the correct one. A probability is assigned to the different states in light of the event's occurrence based on prior experience. An example is that the five-year survival rate for women with Stage II breast cancer is

93% (American Cancer Society 2015a). This means that the mortality rate from breast cancer is 7% because, in Baysian analysis, the sum of the probabilities is 100%.

Although Bayesian analysis is often cited as a basis of clinical decision making, it is not the full story of how clinicians make a diagnosis. As Groopman tells us, "few if any physicians work with this mathematical paradigm. A physical examination begins with the first visual impression in the waiting room, and with the tactile feedback gained by shaking a person's hand. Hypotheses about the diagnosis come to a doctor's mind even before a word of the medical history is spoken" (Groopman 2007, 11–12). The point is that, while empirical evidence plays a key role in how clinicians know in medicine, it is not sufficient for knowledge about disease. Gøtzsche extends the point made by Groopman. He illustrates how the diagnosis of disease entails what he calls "rational knowledge" (Gøtzsche 2007, see title of book). A diagnosis of disease employs a "deductive approach" (Gøtzsche 2007, 61) or "whole to part" thinking, moving from pathophysiological knowledge of disease mechanisms to a particular instance of the disease. When a clinician arrives at a preliminary diagnosis of a disease for a patient who has relevant signs and symptoms and a family history, her reasoning is deductive. It is based on an etiological model of the disease. Her reasoning uses "pattern recognition" (Gøtzsche 2007, 62) to make a diagnosis. A pattern recognition of signs and symptoms can lead a clinician to make a preliminary diagnosis of a disease, one that is followed with a series of tests to confirm or rule out the diagnosis. Her reasoning also uses a "probabilistic diagnosis" (Gøtzsche 2007, 62) based on a statistical calculation of what the condition likely will be prior to ordering tests.

The probabilistic thinking that emerged in the eighteenth century transformed medicine from a body of speculative thinking, "from the play of essence and symptoms" (Foucault 1973 [1963], 105), to one of "calculating" (Foucault 1973 [1963], 89) thinking. It "opened up to [medical] investigation a domain in which each fact, observed, isolated, then compared with a set of facts, could take its place in a whole series of events whose convergence or divergence were in principle measurable" (Foucault 1973 [1963], 97). This freed medical perception from speculative thinking about disease and located it in within "the natural relation between the operation of consciousness and the sign [of disease]" (Foucault 1973 [1963], 95).

Such is a model of empirical thinking set within the framework of rationalist thought. Theory provides a context within which empirical data are set, interpreted, and revised. In turn, empirical data provide the context through which theories are understood and evolve. In the case of breast cancer, theories of breast cancer provide frames for the interpretation of breast cancer data. Breast cancer data provide the context by which theories are developed. Context and content work hand in hand to provide clinical accounts of a disease. For instance, etiological contexts concerning ER+ and breast cancer change and evolve with new knowledge about ER+ status, cellular mutation, and genetic structure and function.

The point is that, although observation and empirical evidence operate in all the examples mentioned, clinicians bring to their diagnosis of breast cancer rational frames of reference through which they "know" breast cancer. So-called rational ways of knowing frame and organize medical thinking and provide a basis for clinical explanations. A major way medicine frames, organizes, and explains empirical and rational evidence for a clinical condition is through causal explanations, a topic that is considered in the next section.

CAUSAL RELATIONS

A dominant way to explain breast cancer is through causal explanations. Causal explanations serve important roles in medicine in explaining why a clinical condition occurs. Ancient Greek philosopher Aristotle (384–322 B.C.E.), who was the son of a physician, held that causality acts as an explanation "for the sake of understanding, and we think that we do not understand a thing until we have acquired the *why* of it" (1969, 29). Modern clinicians, such as Giovanni Morgagni (1682–1771) (1981 [1761]), were particularly interested in developing causal explanations for disease. They voiced criticism of what they called "speculative" accounts of disease which they inherited from the ancient and medieval clinicians. A causal account, they reasoned, would provide an explanation of a disease based on empirical evidence, the means to predict what would happen in the future, and the recipes to guide treatment plans. It is in this spirit that Morgagni says, "the external cause of the disorder is wanting" (1981 [1761], 159). Great strides in developing causal explanations were made by nineteenth- and twentieth-century clinicians, particularly in the areas of infectious disease explanations. One thinks here of Louis Pasteur (1822–1895), who discovered the principles of vaccination, microbial fermentation, and pasteurization, and then developed the first vaccines for rabies and anthrax. One thinks here as well of Robert Gallo (1937–present), one of the discoverers of the human immunodeficiency virus (HIV), the infectious disease agent of acquired immunodeficiency syndrome (AIDS). Isolating what causes bring about what particular clinical conditions leads to successful ways of treating particular clinical conditions.

Causal explanations assist in explaining breast cancer. They forecast the occurrence of breast cancer and guide treatment recommendations and decisions. There are different ways to talk about the cause or etiology of disease and, for purposes of this discussion, consider four senses of the cause of breast cancer. To begin with, there is a sense of clinical causal relation that establishes the necessary relation between the clinical cause and the effect, where the cause is that which brings the effect about and the effect is the disease. This is called "necessary causal relation." Here, "[f]actor X is a *necessary* determinant, if X always precedes Y, which of course does not imply that Y always succeeds X" (Wulff 1981, 56). In the case of breast cancer, breast cancer factor X is a necessary condition of breast cancer (Y) if breast cancer (Y) is always preceded by the breast cancer factor X. For instance, BRCA1 breast cancer is always preceded by a particular gene on chromosome 17 called BRCA1. Here BRCA1 turns off a mechanism in a breast cell that controls cell regulation and growth. In this way, the gene BRCA1 (or mutation on chromosome 17) is a necessary condition for BRCA1 breast cancer, although one can have the gene BRCA1 and yet not express breast cancer, as was the case for actress Angelina Jolie (Jolie 2013, 2015).

A second sense of causal relation is "sufficient" causality. A relation of sufficient causality is less certain than necessary causality but nonetheless provides a sense of what brings breast cancer about. Here, "[f]actor X is a *sufficient* determinant, if X always leads to Y," which "does not imply that Y is always preceded by X" (Wulff 1981, 57). In the case of breast cancer, breast cancer factor X always leads to breast cancer (Y), but there are other paths by which breast cancer can occur. For instance, uncontrolled cellular growth in breast ducts is a sufficient cause for breast cancer. Yet there are other causes of breast cancer. Uncontrolled cellular growth in the lobular or connective tissue of the breast leads to lobular or cribriform carcinoma, respectively.

A third sense of causal relation is referred to as "contributory." In medicine, "[s]ome determinants are neither necessary nor sufficient, but *contributory*. This term is used if some factor X leads to an increased probability of Y, although X does not lead to Y and Y is not always preceded by X" (Wulff 1981, 57). In the case of breast cancer, some breast cancer factor X leads to an increased probability of breast cancer (Y), although the factor X cannot be said to cause the breast cancer (Y) and breast cancer (Y) is not always preceded by the factor X. More specifically, consider the correlation between long-term use of hormone therapy and breast cancer. Taking hormones (and specifically estrogen plus progestin in postmenopausal women) has been correlated with an increased probability of developing breast cancer (Women's Health Initiative 2015), but hormones cannot be said to directly lead to breast cancer and breast cancer is not always preceded by taking hormones. Many women on hormone therapy do not have breast cancer, and many women with breast cancer have never been on hormone therapy. Other factors appear to be necessary and sufficient in the relation between taking hormone therapy and expressing breast cancer.

A fourth sense of causal relation in breast cancer diagnosis is "spurious," or accidental. While breast cancer specialists attempt to avoid making this type of causal relation, it is not difficult to find examples of this type of claim in discussions about breast cancer. Even if the correlation between causal factor X1 and disease (Y) appears genuine, it may not indicate a causal relation if X1 and Y have a common cause (X) or if X1 is a concomitant factor to X in causing Y. Consider a case in which a particular population of women expresses a significant rate of breast cancer. Let's say that these women are Roman Catholic nuns who have been observed to express breast cancer at a higher rate than other women (Olson 2002, 21–24). A tempting conclusion is to say that nunhood causes breast cancer. Our consideration of spurious causal relations reminds us that this is what philosophers call a hasty generalization. A hasty generalization occurs when a faulty generalization is reached by an inductive generalization based on insufficient evidence. Perhaps the previously mentioned population of nuns was more vulnerable to breast cancer at the start because of some other factor or set of factors, such as environmental ones. Perhaps all the nuns were exposed to some kind of environmental toxin that can be shown to correlate with breast cancer. If so, the relationship between being a nun (X1) and breast cancer (Y) may be due to a common cause (X) (called a *confounder*) (Figure 3.1). It is also possible that being a nun (X1) is a *concomitant* factor (X) to the determining factor or factors of breast cancer (Y) (Figure 6.1b). It would be concomitant if a lack of disruption of estrogen production brought about by pregnancy, childbirth, and breast feeding contributes to breast cancer (World Health Organization 2015). In these types of discussions, researchers and clinicians seek to rule out confounding and concomitant factors when determining what causes breast cancer.

As seen here, determining the causal relation between that which brings disease about and the disease occurrence is complex. Various accounts of causal relation span a range of causal claims, including necessary, sufficient, contributory, and spurious. At first , one might think that most causal claims in medicine are necessary or sufficient. But this is not the case. As Gøtzsche says, "the reader of a textbook of medicine may easily get the impression that the causes of many diseases are well known, but when it is stated that one of the many aetiological [causal] factors has been isolated, and when it is stated that the pathogenesis is known, it usually only means that a few links in the chain of events leading to the clinical picture have been discovered" (Gøtzsche 2007, 56). In this case, the presence of BRCA1 guarantees the occurrence of breast cancer.

Applied to breast cancer, at first we may think that BRCA1 is an example of a "necessary" causal relation. But further considerations may lead us to rethink this view. As previously mentioned, there are those who test positive for BRCA1 who do not express breast cancer. This means that something else is going on in confirmed cases of BRCA1 breast cancer. There are contributory factors operating in such cases of breast cancer. While breast cancer specialists aspire to provide explanations of breast cancer that provide necessary or sufficient causal relations, they typically cannot. They cannot because of the complexity of cancer pathways, insufficient data about why breast cancer occurs, and the limits of empirical data in accounts of breast cancer.

Nevertheless, in the twenty-first century, there have been significant studies determining the causal relation between that which brings about breast cancer and the resulting breast cancer. Such determinations have led to significant advances in how breast cancer is treated. In the case of BRCA1, while we do not understand the mechanisms by which BRCA1 develops, we do know that patients have a 55% to 65% chance of expressing breast cancer if the breast tissue is not removed (National Cancer Institute 2015a, 2). In the case of HER2+ and ER+ breast cancers, while we do not know why HER2+ and ER+ breast cancers come about in particular patients, we do know something about how to prevent the progression of these types of breast cancer with Herceptin and Tamoxifen, respectively (National Cancer Institute 2015c). In the case of triple negative breast cancer, while we do not understand how it comes about, we do know that certain treatments are better than others for this aggressive type of breast cancer (National Cancer Institute 2015c). The point is that causal explanations of breast cancer can and do matter, even when they may be incomplete. They matter in terms of decreasing incidence rates of breast cancer in the United States since 2000, with a 7% drop from 2002 to 2003 following the decline in hormone use among women following the findings of the Women's Health Initiative (2015). They matter in terms of death rates from breast cancer since 1989, with the largest decreases being among women younger than 50 (American Cancer Society 2015e).

Causal explanations of breast cancer serve to forecast future occurrences. As feminist philosopher of science Helen Longino (1944–present) tells us, explanation serves as a basis for prediction in the sciences. As Longino puts it, "The purpose of scientific inquiry is not only to describe and catalog, or even explain, that which is present to everyday experience, but to facilitate prediction, intervention, control, or other forms of action on and among the objects in nature" (2002, 124). If we know what brings something about, we can talk better about what will occur. In the case of breast cancer, if we know that X factor brings breast cancer about, we can predict to some extent that, given the presence of X factor, breast cancer may occur. In this way, clinical explanations serve as a basis for clinical prognoses. As we have learned, breast cancer specialists provide contributory causal explanations for breast cancer. It does because medicine as a science offers probabilistic explanations, and not ones that are 100% certain. Tied to this, prognostic indicators will also be less than 100% certain, thereby providing less-than-certain forecasts of the occurrence of breast cancer. While this may come as a relief to those who have been diagnosed with a precancerous condition that may not lead to breast cancer, it wreaks havoc on those who are looking for more certain diagnostics and prognostics for the occurrence of breast cancer. In this light, let us now turn to a topic dear to my journey as a breast cancer patient—the question of how to manage the uncertainty of breast cancer explanations.

COGNITIVE CERTAINTY

The prior discussion of the causes that bring about breast cancer leads to the recognition that breast cancer medicine provides less-than-certain explanations for why breast cancer occurs. This can be a disconcerting thought. When patients go to their doctors, they typically expect a high level of certainty, they expect clear diagnoses, and they hope to be cured, if not relieved of the pain and suffering that brings them to the clinician's attention. In my own experience as a breast cancer patient, I was no different. I certainly hoped for a clear diagnosis of my breast cancer and for answers to my questions about why I expressed breast cancer, what can be done to treat it, and what can be done to prevent its occurrence. It seems to me to date that such expectations are not unreasonable. Since the 1990s, breast cancer clinicians have been increasingly successful in diagnosing and treating breast cancer. They are able to detect more early-stage breast cancer than ever before and distinguish among different types of breast cancer (American Cancer Society 2015e, 1). The have been able to decrease the mortality rates from breast cancer through the use of mammography by 28% since the 1990s (MedLine Plus 2014). Since 2000, breast cancer clinicians have been able to decrease the incidence rate of breast cancer, with a 7% drop from 2002 to 2003 following the decline of hormone use among women (American Cancer Society 2015e). They have been able to develop more targeted treatments for different types of breast cancer and offer treatments that are less disfiguring and have fewer side effects (National Cancer Institute 2014d. There is much to celebrate in breast cancer medicine, so why not hope for diagnoses, prognoses, and treatments that are more certain?

The mindset of expecting certainty in medicine is not new. It represents what philosopher of science Raphael Sassower and physician Michael Grodin call "old notions of certainty—knowing for sure, having precise knowledge about specific events at specific times" (1987, 224). This mindset constitutes an "old notion of certainty" in that it is assumed that medicine as a science would be able to provide certain knowledge. Here one is reminded of early modern clinicians, such as physician Thomas Sydenham (1624–1689) (1753), who, in responding to the need to develop epistemological methods that were free from speculative thought, developed epistemological methods based on what he thought were clear and distinct ideas. The quest for certainty in medicine is not an unreasonable hope in that patients come to health care professionals for knowledge about their health status, knowledge that they cannot easily obtain from home tests. As Sassower and Grodin put it, "[k]knowledge . . . may be of an instrumental nature (to know for a purpose or an end)" (1987, 225). Patients want to know what to do about the signs or symptoms that brought them to the attention of health care professionals. They hope for a clear diagnosis, an effective treatment, and knowledge about why a condition has occurred and how it can be prevented. Similarly, health care professionals hope for a high level of certainty about the diagnosis, treatment, and prognosis of a patient's health status. Anything else would appear to be less than fruitful in medicine.

There are reasons for seeking certainty in diagnostic, prognostic, and treatment claims. A sufficient level of certainty is important in the health care professional–patient relationship for purposes of fostering clear and informed communication. A sufficient level is important for disease nosologies (classifications) and nosographies (descriptions), health insurance coverage, medical negligence accountability, and health policy debates (Sassower and Grodin 1987, 226). In the case of breast cancer, a sufficient level of medical certainty is assumed in the informed consent process in breast cancer medicine. It allows for the development of clear nosologies and nosographies and clear guidelines from insurers about what clinical conditions will be covered and which ones will not. A sufficient level of medical certainty is important in legal matters. It is difficult to hold individuals accountable for the care of breast cancer patients without a sense of what they should be accountable for. A sufficient level of medical certainty is also important in the development of breast cancer health policies and laws. Breast cancer health policies and laws cannot be formulated and implemented well without a sense of what is being addressed.

Yet, as one would expect from the previous analysis, medical certainty is not fully achievable in breast cancer medicine. As Engelhardt says in writing about cancer classifications in general, "[t]here are no unique lines in reality to which the classifications correspond. Rather, the categories are as much created as they are discovered through endowing certain findings with significance" (1996, 220). Recall that even in the case of BRCA1 breast cancer, a mutation on chromosome 17 is a necessary condition, but not a sufficient one for the occurrence of breast cancer. One can have a BRCA1 mutation, yet not have breast cancer. Breast cancer specialists have yet to determine what are the sufficient conditions for BRCA1, and even when they do, medicine will at best be able to provide statistical or probabilistic explanations of breast cancer, and not deductively certain ones. The empirical method used by practitioners limits their ability to provide deductive certainty about disease. It limits it because health care practitioners and researchers can at best speak in inductive probabilities about clinical diagnosis, prognosis, and treatment. Gone are the days of an "old notion of "certainty" in medicine.

According to Sassower and Grodin, "modern notions of certainty" were developed alongside the development of the theory of probability. Here a theory of probability accounts for the extent to which something is probable and the likelihood of something happening or being the case (Burger and Starbird 2005). Probability is determined in mathematics by the ratio of the favorable cases to the whole number of cases. It gives meaningful numerical value to the occurrence of an outcome or set of events when such an outcome or set of events is uncertain or unknown. This modern notion of certainty arose in the seventeenth century with mathematician and attorney Pierre Fermet (1601–1665) and mathematician and philosopher Blaise Pascal (1623–1662) in the context of trying to understand the phenomenon of gambling (Burger and Starbird 2005). Speaking of calculations of probabilities, Pascal says, "The uncertainty of the gain is proportional to the certainty of the stake according to the proportion of the chances of gain and loss" (Pascal 1910 [1670], Ch. 16). The science of probabilities continues to develop and provide insight into how decisions are made in light of risk assessment (Bell, Raiffa, and Tversky 1988).

Modern notions of certainty, or uncertainty, pervade medical thinking. As Foucault reminds us, the shift from speculative to probabilistic thinking in medicine opens up the investigation of signs, symptoms, pathology, and etiology of disease in a way that is measurable though probabilistic calculations. In this shift, the plurality of observations is "no longer simply a contradiction or confirmation, but a progressive, theoretically endless convergence" (Foucault 1973 [1963], 97). In this shift, medicine discovers "that uncertainty may be treated, analytically, as the sum of a certain number of isolatable degrees of certainty" (Foucault 1973 [1963], 97) that are "capable of rigorous calculation" (Foucault 1973 [1963], 97). In the end, medical knowledge gains in certainty "in relation to the number of cases examined" (Foucault 1973 [1963], 101). As Foucault says, "medical certainty is based not on the *completely observed individuality* but on the *completely scanned multiplicity of individual facts*" (Foucault 1973 [1963], 101). On this view, medical uncertainty is measureable, changing, and contextual.

The uncertainty that pervades medical knowledge comes in various kinds. Groopman, drawing on the work of University of Pennsylvania sociologist Renée Fox (1928–present), identifies three basic senses of uncertainty that are evident in medicine. "The first results from incomplete or imperfect mastery of available knowledge. No one can have at his command all skills and all knowledge of the lore of medicine" (Groopman 2007, 152). In medicine, where there is a huge body of knowledge, it is unreasonable to expect that any single individual has command of all knowledge on a certain disease. In the case of breast cancer, there are a wide range of specialties in breast cancer care, including breast cancer surgeons, medical and nursing oncologists, radiation oncologists, breast reconstructive surgeons, psychologists, and social workers. Specialists from one area of breast cancer care are not expected to be experts in other areas (although they are expected to know when to refer a patient to another specialist for care). Further, any individual breast cancer specialist at any one time cannot be expected to have all the relevant knowledge of a particular aspect of breast cancer care (although they are expected to be "competent" in their area of expertise).

"The second [sense of uncertainty in medicine] depends on limitations in current medical knowledge. There are innumerable questions to which no physician, however well trained, can provide answers" (Groopman 2007, 152). As we have learned, medical knowledge is limited. It is limited because the kind of empirical or inductive methodology that is used in medicine is only able to make claims that are temporary. Claims are temporary because medical knowledge changes and evolves. In its evolution, there are numerous claims that are advanced, numerous questions that go unanswered, and numerous views that are changed with the rise of new knowledge. Since the time of Hippocrates, there have been numerous accounts of why breast cancer occurs, ranging from anatomical to physiological to environmental to biomolecular. There are numerous claims that have been changed, such as "an increase in black bile causes cancer," "hormone therapy does not contribute to breast cancer," "a radical mastectomy is the best treatment for breast cancer," and "there's no treatment that medicine can offer to late-stage breast cancer patients." There are numerous questions that go unanswered, such as why certain types of Stage 0 breast cancer develop into later stages of breast cancer and others do not, and why certain treatments for breast cancer work for some women and not for others. Medical knowledge is limited to inductive claims, which change and evolve with the rise of new evidence and theoretical shifts in our understanding.

A third source of uncertainty derives from the first two: "this is the difficulty in distinguishing between personal ignorance or ineptitude and the limitations of present medical knowledge" (Groopman 2007, 152). Given the rate at which knowledge evolves in medicine, it makes sense that managing the boundaries between one's own lack of knowledge and the lack of knowledge in medicine can challenge even the best informed breast cancer specialist. Consider trying to keep up with knowledge about breast cancer. Prior to 1900, the wide variety of kinds of breast cancer were not known. Prior to 1980, the HER2 gene, the product of which is found on the cell membrane and correlates with certain types of breast cancer, was not known (Mukherjee 2010, 410). Prior to 1990, the genes for BRCA1 and BRCA2 had yet to be discovered. Prior to 2000, adjuvant therapy for breast cancer was not available. Prior to 2010, immunotherapy for breast cancer was a distant hope. Coming to terms with what one knows and does not know, and what clinical practitioners know and do not know, can challenge even the most informed clinician and patient.

Another source of uncertainty has to do with the kind of evidence medicine provides. Medicine provides for the most part group data for disease occurrence, etiology, prognosis, and treatment responses. It does this because medicine relies on populations of research subjects and studies about how disease comes about, how patient populations respond to treatment, and what happens in the future to those who have or have not been on a certain treatment regime. But population data are just population data, and not data about an individual patient and her disease condition. In the case of breast cancer, population data can tell an individual patient about how many women express a certain kind of breast cancer and what happens to them while on a certain treatment plan, but it does not tell a patient why she has a particular kind of breast cancer and what will happen to her when she proceeds with a certain treatment plan. Recalling a personal story shared earlier in the Preface, while a clinical research report may say that a breast cancer patient who has been diagnosed with a certain kind of breast cancer and undergone a certain kind of treatment will have a particular probability of survival based on population data, it does not tell *me* which group I will be in the future. It does not tell me about whether I will live or die in the next five years of my life. As physician and historian of medicine Barron H. Lerner tells us, "one of the major challenges for the upcoming decades is reconciling these public discussions [about diagnosis, prognosis, and treatment] which generally rely on population data, to the dilemma faced by individual women" (Lerner 2001, 241) in choosing one set of options over the other in breast cancer care.

The recognition of the "limitations of human reason" as found in these various senses of certainty leads to "tension," as Engelhardt puts it, "between the universal aspirations of knowers and the particular context in which real individuals actually know and frame explanations" (1996, 218). Stating what is known and not known in an academic discussion may not give us pause; stating what is known and not known in medicine and in the context of treating breast cancer patients may give pause to patients and their clinicians. Clinicians are not trained to be comfortable with clinical diagnoses, prognoses, and treatments that are uncertain. Patients are even less comfortable with such uncertainty, especially in high-risk clinical contexts. Such is the condition of knowing that we do not know with 100% certainty, knowing enough about what we do not know, and being constrained by a methodology for knowing that cannot guarantee certainty. Such is the human epistemological condition of limited knowledge.

COGNITIVE SKEPTICISM

The problem of cognitive certainty, or rather uncertainty, raises a related philosophical problem called the *problem of skepticism*. Generally put, skepticism is the tendency to deny or doubt claims. In the case of breast cancer, given we cannot know breast cancer with 100% certainty, it is tempting to conclude that we or "they" know nothing about breast cancer. It is tempting to conclude that breast cancer specialists make up what they say and patients believe what they are told. We all know that one hundred years from now our ancestors will be astonished at how we understood and treated breast cancer. Nevertheless, we can say that we know something about breast cancer, although such knowledge is not immune from a healthy sense of skepticism.

We start by considering different kinds of skepticism, three of which are mentioned here, along with their application in the context of knowing breast cancer. At one end of the spectrum is what I call *absolute skepticism.* Absolute skepticism doubts the possibility of knowledge altogether. It is associated with the thinking of Sophist philosopher Gorgias Leoritini (approximately 525 B.C.E.), who holds that nothing whatsoever is certain. Gorgias' position can be expressed in four theses: (1) nothing exists; (2) if something did exist, we could never know it; (3) if we could know it, we could never express it to others; and (4) if we cannot express it to others, it cannot be understood (Consigny 2001). It is safe to say that *absolute skepticism* is not compatible with knowing disease in general, and breast cancer in particular. Individuals who are absolute skeptics would likely not seek medical advice and, if they do, they would not follow through with their clinical appointments and recommendations (Ericksen 2008, 34). Taken to the extreme, absolute skeptics deny the legitimacy of any and all knowledge, including medical knowledge. Because of this, absolute skepticism is not relevant in this inquiry, one that holds that we can know something about breast cancer.

Second, *philosophical skepticism* doubts most cherished claims, such as taken-for-granted or foundational claims about knowledge. Philosophical skepticism differs from absolute skepticism in that philosophical skepticism does not doubt the possibility of knowledge as does absolute skepticism. It is associated with the thinking of the Ancient Greek skeptics who believed that knowledge of the world was only an approximation or opinion (Annas and Barnes 1985). Ancient Greek philosopher Diogenes Laertius (approximately third century C.E.) summarizes the way to skepticism: "The Skeptics . . . were constantly engaged in overthrowing the dogmas of all schools, but enunciated none themselves; and though they would go so far as to bring forward and expound the dogmas of the others, they themselves laid down nothing definitely, not even the laying down of nothing" (1925, IX, 75). Note here that Diogenes critiques the dogma, knowledge, and ideas of the time without supplying an alternative account. This is a strategy common to philosophical skeptics because any alternative account would be susceptible to skeptical analysis and therefore would not be able to be accepted as truth.

Here, I think, there is application in our discussion of breast cancer. Sociologist Julia Ericksen, a breast cancer survivor herself, details the case of those who are skeptical of knowledge in breast cancer medicine. She refers to such patients as "alternative experts" (2008, 34) and "religious experts" (2008, 34). Alternative experts use "alternative practices to supplement or replace some aspects of standard medical treatment" (Ericksen 2008, 34). Religious responders usually follow their doctors' orders but believe "they must put their ultimate faith in God and not in doctors" (Ericksen 2008, 34). They use "prayers . . . to help them safely through their illness" (Ericksen 2008, 34). Together, alternative and religious responders doubt allopathic medicine's approach to breast cancer, but not all approaches. They hold that alternative care (e.g., dietary changes) or religious intervention (e.g., prayer) is the best treatment for breast cancer. At the same time, if they believe that alternative or religious intervention is insufficient, they can and will turn to allopathic medicine for clinical help, continually submitting claims about breast cancer to philosophical skeptical analysis.

Third, *common-sense skepticism* is a healthy corrective to gullibility, superstition, and prejudice. It is an antidote to intellectual arrogance and presumption, and an offspring of philosophical skepticism. It is commonly found in philosophical writings and often expressed in terms of doubts or objections to claims that are made. In the case of breast cancer, a healthy sense of uncertainty allows the breast cancer specialist to step off of her or his pedestal, engage in dialogue with the patient, and change the course of diagnosis and treatment, assuming there is good reason to do so. A healthy sense of skepticism plays an important role in medicine. As Groopman puts it:

> Does acknowledging uncertainty undermine a patient's sense of hope and confidence in the physician and the proposed therapy? Paradoxically, taking uncertainty into account can enhance a physician's therapeutic effectiveness, because it demonstrates his [or her] honesty, his [or her] willingness to be more engaged with his [or her] patients, his [or her] commitment to the reality of the situation rather than resorting to evasion, half-truth, and even lies. And it makes it easier for the doctor to change course if the first strategy fails, to keep trying. (2007, 155)

Given this, and as Groopman says, "[u]ncertainty sometimes is essential for success" (2007, 155). Perhaps, better put, recognizing uncertainty is essential for success because health care professionals and their patients operate within a world of uncertain medical knowledge.

One might say as well that it is appropriate for a patient to have a healthy sense of skepticism about clinical diagnosis, prognosis, and treatment. From the patient's standpoint, it is appropriate to ask questions, obtain second opinions, and be informed, a practice endorsed by the American Cancer Society (2015e), the National Cancer Institute (2014d), and numerous breast cancer organizations (e.g., Susan G. Komen, Avon Foundation, breastcancer.org). A healthy sense of skepticism is to be seen as a check against unsubstantiated claims and interventions. It also serves an important role in guiding decision making in the context of incomplete and evolving information about breast cancer.

In addition, a healthy sense of skepticism combats the tendency to think that the claims made in medicine are deterministic. *Determinism* is the view that every event has a cause (Pojman and Feiser 2011, 623). Here the view is that, for every event, there is some antecedent state, related in such a way that it would break a law of nature for this antecedent state to exist and the event not to happen. At first, it is tempting to think that diagnostic claims in medicine provide deterministic accounts of breast cancer. That is, it is tempting to think that a positive test for BRCA1 forecasts the inevitability of breast cancer. But it does not. As discussed in the section on "Causal Relations", having BRCA1 does not determine that one will express breast cancer, although there is a greater likelihood that a patient diagnosed with BRCA1 will express breast cancer than if one does not have BRCA1. The same can be said for calcification, or calcium deposits, in the breast tissue. A positive diagnosis of calcification in the breast does not predict that one will express breast cancer, but it increases the likelihood that something abnormal is going on in the breast. Because data in medicine are limited and uncertain, it is important to be reminded that the claims made in breast cancer care are not deterministic. Breast cancer specialists speak in probabilities that are not 100% certain. They do this because they cannot do otherwise given their reliance on an empirical methodology that limits the claims that can be made. While such a conclusion might be lamented, it may bring some cognitive relief to those who seek certainty in breast cancer medicine. It brings about cognitive relief because one gives up the unachievable hope of certainty and accepts the limits of medical claims and the challenges that come with managing the cognitive uncertainties of medical diagnosis, prognosis, and treatment.

CLOSING

The question "How is breast cancer explained?" raises a host of epistemological issues, including ones about the methodology we use to explain breast cancer, the etiology of breast cancer, and the extent to which our explanations of breast cancer provide clinical certainty. What we find is that, initially, it appears that clinical practictioners and researchers explain breast cancer using an empirical methodology that provides necessary or sufficient causal relations that grant certain knowledge. But things are not as simple as one would initially think. Upon reflection, explanations of breast cancer depend on the methodology through which they are seen and interpreted. Causal accounts of breast cancer are not typically modeled on a simple cause-and-effect relation. Given this, clinicians and patients alike must face the recognition that clinical explanations of breast cancer are probabilistic and a healthy sense of skepticism provides a check against an idealized sense of knowledge about breast cancer. This does not mean that anything goes, for the clinical endeavors of interpreting evidence and advancing etiological accounts based on a large number of clinical cases gives us an understanding of and treatments for breast cancer.

Chapter 4

How Is Breast Cancer Evaluated?

PERSONAL MUSINGS

As a breast cancer patient, I often thought about how the descriptions and explanations for my breast cancer were evaluated in terms of the actions that clinical practitioners deemed appropriate to take place. The Tuesday afternoon I received word from my breast cancer surgeon that she was "surprised" by the surgical biopsy results, I recognized that my life would never be the same. I knew that the results of my diagnostic tests would set up a series of clinical actions that would change my life. At each of my clinical appointments, clinicians and I waded through the incoming clinical information. I prepared myself for the implications of my newly acquired clinical labels. I pondered how I would come to terms with the "c" word. I threw myself into the breast cancer literature as recommended by my doctors and tried to understand my diagnosis and treatment options. Today I ask myself, how did the diagnosis of my breast cancer serve as a guide for treatment? How did it point in the direction of recommending a certain series of interventions? Put philosophically, how did the diagnosis set up a series of "oughts"? Did the diagnosis and treatment recommendations appeal to a set of clinical values? If so, what kind of values? To what extent did the diagnosis and treatment recommendations reflect evaluations of human function, the achievement of specified ends or goals, ideals of beauty, and moral considerations? And whose values were they? Did they reflect my own and my clinician's values?

Given the role the insurance and law play in guiding breast cancer treatment, how did they reflect my insurer's and my government's values as well? Given the role values play in clinical descriptions and explanations, to what extent did the role of values in understanding my breast cancer make its diagnosis and treatment even more less-than-certain than they would be if clinical descriptions and explanations were value-neutral? To what extent could I, or perhaps should I, be skeptical of the evaluations of my breast cancer? Herein lay the basis for some of my philosophical reflections on the evaluative dimension of breast cancer and the axiological or evaluative issues concerning normativism, kinds of clinical values in medical thinking, and value certainty in the context of treating breast cancer. In this chapter, and in the context of evaluating breast cancer, neutralism is contrasted with normativism, kinds of clinical values are distinguished, value objectivism is contrasted with value subjectivism, and different notions of value skepticism are considered (see Figure 4.1).

(a)

value neutralism	normativism
(the view that values play no role in claims of facts)	(the view that values play a role in claims of fact)

(b)

functional	instrumental	aesthetic	moral
(value of operation)	(value of means to end)	(value of beauty)	(value of praiseworthiness or blameworthiness)

(c)

value objectivism	value subjectivism
(the view that values are discovered	(the view that values are created and personal and shared)

(d)

absolute value skepticism	philosophical value skepticism	common-sense value skepticism
(doubts possibility of value statement altogether)	(doubts most cherished value claims and assumptions)	(questions value claims and assumptions)

Figure 4.1 Axiological issues.

VALUE NEUTRALISM

Initially, there is a tendency in medicine to view breast cancer as a value-neutral phenomenon or "fact." By "fact," I mean an empirically verifiable statement that correlates with an object or state of affairs that exists. A fact is typically contrasted with a "value" (from the Latin root *valere,* meaning "to be of worth"). Here "value" is used broadly to refer to a sign of worth or assessment of comparison. Given that medicine relies on scientific methods to investigate and determine the nature of and treatment for particular diseases, medicine is typically seen to deliver value-neutral accounts of disease and its treatment. On this view, disease is a "fact" that is arrived at through scientific investigation and is tested through verification or falsification by others.

In the case of breast cancer, the TNM staging system of classification for breast cancer is based on empirically verifiable accounts of statistical normality and the biological function of breast cells. Let's take a look at a supposedly value-free account of ductal breast cancer as mapped out by the TNM staging system of classification. As a reminder, ductal breast carcinoma is found in the milk ducts and affects close to 70% of all breast cancer patients. The American Joint Committee on Cancer has developed general guidelines for treatment for each of the TNM stages of ductal carcinoma of the breast (see Figure 4.2; for more detail, see American Joint Committee on Cancer 2017). As stipulated here, "primary treatment" targets the source or process of a clinical condition and "secondary treatment" increases the chance of long-term survival from the clinical condition. A key to the acronyms and terms follows the figure.

As seen here, and as one would expect, treatment options depend on the stage of the breast cancer. In addition, they depend on the size of the breast, biomarker tests, whether the patient has gone through

Stage of Ductal Carcinoma	Primary and Secondary Treatment Options
Stage 0 (Tis, N0, M0)	Primary: a lumpectomy or total mastectomy with no removal of lymph nodes; radiation treatment may be recommended Secondary: adjuvant or targeted therapy (if applicable)
Stage IA (T1, N0, M0) Stage IB (TO, N1mi, MO, or T1, N1mi, MO) Stage IIA (T0, N1, M0 or T1, N1, M0 or T2, N0, M0) Stage IIB (T2, N1, M0 or T3, N0, M0)	Primary: lumpectomy or mastectomy, with sentinel lymph node biopsy (and axillary lymph node removal if needed); radiation Secondary: adjuvant or targeted therapy (if applicable)
Stage IIA (T2, N0, M0) Stage IIB (T2, N1, M0 or T3, N0, M0) Stage IIIA (T0, N2, M0 or T1, N2, M0 or T2, N2, M0 or T3, N1, M0 or T3, N2, M0)	Primary: lumpectomy or mastectomy, with sentinel lymph node biopsy (and axillary lymph node removal if needed); chemotherapy, radiation Secondary: adjuvant or targeted therapy and chemotherapy
Stage IIIB (T4, N0, M0 or T4, N1,M0 or T4, N2, M0) Stage IIIC (any T, N3, M0)	Primary: lumpectomy or mastectomy, with sentinel lymph node biopsy (and axillary lymph node removal if needed); chemotherapy, radiation Secondary: adjuvant or targeted therapy and chemotherapy
Stage IV (any T, any N, M1)	Primary: lumpectomy or mastectomy, with sentinel lymph node biopsy (and axillary lymph node removal if needed); systematic treatment with chemotherapy and adjuvant therapy

Key: T = tumor size (see Chapter 3 for explanation)
is = in situ (tumor that has not grown into breast tissue)
N = number of lymph nodes (see Chapter 3 for explanation)
M = metastasis or number of sites to which cancer has spread (see Chapter 3. for explanation)
mi = micrometastases (deposits of molecularly or microscopically detected tumor cells in nodal tissue less than 0.2 mm)
lumpectomy = removal of a tumor from the breast
mastectomy = removal of the entire breast tissue (and, perhaps, a number of axillary lymph nodes, the lining over the chest muscles, and underlying chest muscles)
radiation = use of high-energy rays to destroy cancer cells
adjuvant therapy = pharmaceutical agent that eradicates the reservoir of malignancy left in the body after primary therapy and increases the chance of long-term survival
targeted therapy = pharmaceutical agent that is directed at a specific type of cancer as determined by biomarker or molecular tests
sentinel node biopsy = removal of initial lymph node
axillary node removal = removal of lymph nodes under the arm
chemotherapy = pharmaceutical agent that targets cancer cells

Figure 4.2 Disease treatment: primary and secondary options for ductal carcinoma of the breast.

menopause, the patient's general health, the patient's choice, and other considerations raised by the breast cancer specialist (Patient Resource 2012, 22). Treatment seeks to remove (as in the case of a mastectomy), stop (as in the case of chemotherapy), or prevent (as in the case of adjuvant therapy) ductal breast carcinoma. In each case, treatment targets what is known through clinical observation and patient testing, whether this is a sign, symptom, etiology, biological process, or family history of breast cancer. In this way, our understanding of breast cancer appears to be value-neutral and based on the "facts" of the clinical case.

A *value-neutralist* account of reality gained popularity in science and philosophy in the beginning of the twentieth century. It was particularly supported by those called the "logical positivists," who held that knowledge is based on statements that can be verified through empirical observation. Logical positivists heralded the idea that "certain criteria must be fulfilled to warrant calling a body of knowledge science" (Sassower and Cutter 2007, 72). The criteria of empirical testability, including verification or falsification, was particularly important to the logical positivists. Value neutralism is associated with the thinking of British philosopher A. J. Ayer (1910–1989) (1952). Ayer maintains that a proposition is meaningful if it is either analytic (i.e., A = A) or empirically verifiable (i.e., A = B). That is, a proposition is meaningful if it is true as a matter of mathematical definition or it can be supported by empirical evidence. As he says about value propositions, "in so far as statements of values are significant, they are ordinary 'scientific' statements; and that in so far as they are not scientific, they are not in the literal sense significant, but are simply expressions of emotion which can be neither true nor false" (Ayer 1952, 102–103). For Ayer, "significant" statements of values can be empirically supported, and therefore they reduce to statements of "facts." "Nonsignificant" ones cannot be empirically supported; they are expressions of emotions and not able to provide claims that are verifiable or falsifiable.

Philosopher of medicine Christopher Boorse defends a *value-neutral account of disease* in medicine. He defines disease as follows:

> 1. The *reference class* is a natural class of organisms of uniform functional design; specifically, an age group of a sex of a species.
> 2. A *normal function* of a part or process within members of the reference class is a statistically typical contribution by it to their individual survival and reproduction. . . .
> 3. A *disease* is a type of internal state which is either an impairment of normal functional ability, i.e., a reduction of one or more functional abilities below typical efficiency, or a limitation on functional ability caused by environmental agents.
> 4. *Health* is the absence of disease. (Boorse 1977, 562, 567; see also Boorse 1997, 7–8)

For Boorse, the reference class relies on uniform design found in an age group of a sex of a species. By "normal functional ability," Boorse means "the readiness of an internal part to perform all its normal functions on typical occasions with at least typical efficiency" (Boorse 1997, 8). On Boorse's account, to call something a disease involves a claim about the abnormal functional ability of some bodily system and how it compromises individual survival and reproduction in a specified population group. To distinguish between health and disease, or the normal and the pathological, medicine relies on a reference class and an empirically verifiable account of statistical normality and biological function (Boorse 1997, 8). This biostatistical view of disease entails the view that an empirically verifiable account of statistical normality and biological function of the organism correlates with specified treatment recommendations that are shown through clinical investigation and experience to work. At least, that appears to be the view previously mapped out in the stages of ductal breast cancer.

But disease is not as value-neutral as one might think. Again, this may come as a surprise to clinicians and patients alike. Consider that disease is understood, appreciated, and seen through a set of evaluations. As Engelhardt puts it, "[t]o see a phenomenon as a disease, illness, or disability is to see something wrong with it" (1996, 197). In the world of medicine, "[p]roblems stand out as problems for medicine because they are disvalued. They are seen as pathological. They are associated with pathos and suffering, and suffering is judged, all else being equal, to have a disvalue" (Engelhardt 1996, 203–204). Disease is experienced as "failures to achieve an expected state, a state held to be proper to the person afflicted" (Engelhardt 1996, 197). This may be a failure to achieve an expected level of function or ability, an expected level of freedom from pain and suffering, a realization of human form or grace, and/or a state expected by a patient. In other words, the "facts" of disease are in some sense evaluative; they are not neutral.

In the case of breast cancer, breast cancer is not simply a classification or taxonomy for theoretical purposes. It serves to guide clinical actions for purpose of treating patients. As the American Joint Committee on Cancer says, "[t]he primary objective of TNM staging . . . [is] to provide a standard nomenclature for prognoses of patients with newly diagnosed breast cancer and its main clinical utility . . . [is] to prevent apparently futile therapy in those patients who were destined to die rapidly in spite of aggressive local treatments" (2010, 422–423). In other words, a major role of the TNM staging system of classification for breast cancer is to determine the diagnosis of breast cancer for purposes of forecasting its future development and guiding plans for treatment, including the withholding of futile treatment for late-stage breast cancers. Such purposes are evaluative; they designate certain conditions and actions as worthy or significant in order to reach certain ends and achieve certain values.

On this view, breast cancer is a dysfunction of breast cells. It is a failure of breast cells to divide and multiple properly. It is a deviation that occurs in biological pathways that organize and control life processes that contribute to what we call "health." In late-stage breast cancer, breast cancer undermines an expected level of freedom from pain and suffering and, as the breast darkens and puckers, it undermines an expected level of form and grace of the body. If left untreated, breast cancer can lead to premature death and the end of a life that would have otherwise lived. In the end, breast cancer can undermine human choice to live in a certain way, or to live at all, and can harm patients through the pain and suffering that occurs as a result of breast cancer. In this way, breast cancer is not a simple fact; it is an evaluation involving judgments concerning what and how we assign worth or praise, and what we consider "bad" or "harmful." It is an evaluation of what needs to be removed, stopped, or prevented. It is a judgment that a certain phenomenon of human existence is a problem in and for life.

The view that a "factual" phenomenon is evaluative is known in philosophy as *normativism*. Normativism is the view that facts function to some extent as norms or guidelines prescribing certain actions and views. Normativism arises out of a reaction to early-twentieth-century logical positivists or so-called neutralists who held that logical or empirical facts are the sole basis of knowledge. Disagreement between neutralists and normativists typically focus on two points: (1) on the way in which facts are understood and (2) on the way values serve as a basis for justification (Beauchamp

1982, 337ff). With regard to the first (1), philosopher Philippa Foot (1920–2010) holds that the descriptive or factual component of meaning and the prescriptive or value component of meaning are not so separable. Words that are generally used to describe objects also require by their logic that one takes a positive or negative attitude or viewpoint whenever one uses them (Foot, in Warnock 1967, 67–68). Consider, for instance, disease. The notion of "disease" (from the Old French roots *des*, meaning "lack," and *aise*, meaning "ease") entails factual claims about biological functioning but also an attitude about such biological functioning. The attitude reflects that, all things being equal, disease is bad because it is harmful to patients and treating disease is good. With regard to the second (2), Foot holds that there are shared criteria for evaluating things. As she says, "not just anything can function as a criterion for moral justification" (Foot, in Warnock 1967, 67). For a claim to function as a criterion for moral justification it must lie "somewhere within the general area of concern with the welfare of human beings" (Foot, in Warnock 1967, 67–68). In the case of disease, disease serves as a justification for intervention for purposes of minimizing patient harm and enhancing patient welfare. It targets what needs to take place in order to remove that which is harmful. It is the focus of change, if not elimination, for purposes of enhancing patient welfare. In the end, facts and values are not as separate and distinct as one might think.

In philosophy of medicine, the view that disease carries descriptive as well as evaluative dimensions is known as *disease normativism*. On this view, disease is not just a fact; it is also an evaluative notion, which tells us what is significant and why (Cutter 2003, 80). As an evaluative notion, it serves as a treatment warrant, a reason or justification to intervene. It is a particular state of affairs deemed in medicine to be harmful and in need of change or control. As physician and philosopher of medicine Lawrie Reznek says about the harm of disease, "A has a disease P if and only if P is an abnormal bodily/

mental process that harms standard members of A's species in standard circumstances" (1987, 162). Further, "[s]omeone in state S has received harm if and only if he [or she] is worse off in state S than he [or she] was before he [or she] was in state S" (1987, 136). Here, harm consists "in the malfunction of systems worth having, or the frustration of worthwhile pleasures, or the frustrations of worthwhile desires" (1987, 152). Typically, harm is judged, all else being equal, to have a disvalue.

In the case of breast cancer, breast cancer is not just a fact. Breast cancer or *karkinos* is not just a biological tumor; it is an evaluation in medicine that tells us that something is significant and worth changing. It is a particular state of affairs of the breast and of the body that is spreading (like a crab) and needs to be changed in order to attain certain goals. Here a goal in oncology (from the Greek roots, *oncos,* meaning "swelling" or "mass," and *logy,* meaning "study of") is to reduce harm because breast cancer is harmful, or potentially harmful, to life. Breast cancer is a warrant in medicine to set up a series of responses focused on curing, if not preventing or managing, the condition. It is a warrant in medicine that serves as a basis for choices about what actions ought to take place, at what time, and in what order. On this view, breast cancer is understood in terms of a range of clinical values, a topic that is next explored.

CLINICAL VALUES

Let's take a closer look at the kinds of values that frame our understanding of breast cancer. As previously mentioned, the term "value" comes from the Latin *valere,* which means "to be worth." Here "value" can mean at least two different senses. It can be distinguished from "fact" as well as a "disvalue." On the former distinction, "value" is a statement of evaluation of good or lack of good, or praiseworthiness and blameworthiness. Such a statement of evaluation contrasts with "fact," which refers to "a thing done," or a "reality of existence," that is to something that has "really" occurred or is "actually the case." On the latter distinction, "value" refers to a positive judgment and contrasts with "disvalue," a negative judgment (Pojman and Feiser 2011, 107). The discussion of value that follows focuses on the former understanding of "value" as a statement of evaluation that is in contrast to a "fact." While many distinctions may be drawn, I will focus on four kinds of values that are evident in our understanding of breast cancer, namely, functional, instrumental, aesthetic, and moral values (Engelhardt 1996, ch. 5; Cutter 2003, ch. 6; Cutter 2012, ch. 3).

To begin with, *functional values* are at play in our understanding and treatment of breast cancer. Functional values assign worth or significance to how something functions. Since the measure of an organ (e.g., breast), tissue (e.g., connective), or cell (e.g., ductal) is highly uninformative, a clinician has no idea from just one single case what range of data can count as "normal" function and what can count as "abnormal" function. Group or population data are taken into account. What is needed is a range of measurements for a function for a range of populations that clinicians and researchers determine to be "normal" and "abnormal." Here "normal" can have a number of senses: (1) statistical, (2) average or mean, (3) typical or expectable, (4) conducive to the survival of the species, (5) innocuous or harmless, (6) commonly aspired to, or (7) excellent in its class (Murphy 1976, 117–133). Consider a ductal breast cell. A ductal breast cell can be "normal" in the sense of what a "typical" breast cell looks like and how it functions. It can be what an "average" or typical ductal breast cell looks like and how it functions. It can refer to the type of ductal cell that is "conducive to the survival of the species," "harmless" to individual life, "commonly aspired to," or "excellent in its class." These measures are not value-neutral; they turn on a set of determinations regarding what structure or function is statistical, average, or typical for the cell. The judgment about what is statistical, average, or typical is based on views about how the organ, tissue, or cell functions in order to carry out the actions that contribute to the survival of the individual, to reduce harm to the patient, and to contribute to the survival of the species. In all cases, the structure and function of the individual cell is compared to others and determinations regarding what is "normal" versus "abnormal" are made in light of thresholds that are seen to be significant in light of the context.

In the case of breast cancer, breast cancer specialists make determinations about what constitutes normal versus abnormal cell structure and function in the breast. Mutated ductal breast cells are considered abnormal. They grow at a certain rate, say doubling in several years ("Diagnosing Breast Cancer" 2014), which means a cell can take six to eight years to grow a one centimeter size lump found on a mammogram, the general point at which the breast lump can be seen. There is a point at which mutated breast cells may become "suspicious" or "precancerous," "hyperplasia" or "atypical." There is a point at which they may become "carcinoma in situ" (i.e., not yet invading the breast tissue) and "cancerous" (i.e., that which invades the breast cancer tissue). There is a point at which they may become "micrometastatic" (i.e., that which spreads in minimal ways from its original location to other sites in the body) and "metastatic" (i.e., that which spreads from one part of the body to another). A determination of what is normal versus abnormal breast cell structure and function (and the degrees between) appeals to clinical evidence (e.g., clinical observations), scientific evidence (e.g., pathological observations), and value judgments that breast cell growth is abnormal and ought to be attended to. The determination of dysfunctionality involves a value judgment that the cells are not functioning properly and, if left alone, there will be continued dysfunction that can lead to problems. Such judgments turn on assessments of what constitutes normal structure and function among a number of like cases and the observation that variations from such structure and function lead to clinical problems and something ought to be done about it.

In addition to functional values, *instrumental values* frame our understanding and treatment of breast cancer. Instrumental values assign worth or significance to meeting certain ends or goals. In developing classifications for diseases, for instance, clinicians and researchers are concerned to minimize transaction and opportunity costs to the patient and related parties (Cutter 1992, 99). By minimizing transaction and opportunity costs, I mean reducing market and quality of life costs brought about by having a disease and undergoing treatment. If standards for treatment are too lax, this may unduly increase the financial, social, and personal burdens for patients as well as society at large. However, if standards for treatment are too strict, this may result in burdens in terms of lack of treatment and loss of lives. In such situations, medicine attempts to adopt a prudent balancing of the transaction and opportunity costs related to over- and underdiagnosis, as well as over- and undertreatment, in order to maximize benefits and minimize harm to patients and related parties.

In the case of breast cancer, the number of stages of cancer selected presuppose cost–benefit calculations and understandings of prudent actions that have direct implications for the ways breast cancer patients are diagnosed and treated. In 2003, for instance, the American Joint Committee on Cancer recommended that the N (node) category of the TMN (tumor, metastases, node) cancer staging system of classification be changed from one to three categories based on the number of axillary (i.e., under the arm) lymph nodes that are affected by cancer (American Joint Committee on Cancer 2010, 423). This change came about because of a clinical shift in understanding the function of lymph nodes in determining the extensiveness of breast cancer and the need for more specific diagnoses and treatments of breast cancer in order to reduce morbidity and mortality. Another example is drawn from the debate regarding how often to screen for breast cancer, a topic that will be covered further in Chapter 7.

Along with functional and instrumental values, *aesthetic values* operate in our understanding and treatment of breast cancer. Aesthetic values assign worth or significance to that which is found to be pleasing, artful, or beautiful. Aesthetic judgments enter into how we understand disease in the sense that ideal states of physiological, anatomical, and psychological function presume a level of human ability, form, movement, and grace (Khushf 1997; Engelhardt 1981 [1975], 125). Judgments in medicine are aesthetic in that form and function are typically considered beautiful, and deformity and dysfunction ugly. Deformity and dysfunction are typically disvalued because they fail to meet ideals of symmetry, coherence, and visualization, as determined by patients, health care providers, and members of society.

Aesthetic judgments are particularly evident in how we understand and treat breast cancer. To begin with, breast cancer is ugly. A lump in the breast typically undermines the beauty and symmetry of the breast. If left untreated, breast cancer leads to a darkening and puckering of the breast skin, which is typically considered aesthetically displeasing, especially in cultures that place significant value on the beauty of the breast and its role in how sexuality, women's bodies, and reproductive fitness are viewed. If further left untreated, breast cancer can lead to oozing from the breast. Judgments that something is displeasing, unaesthetic, or ugly typically elicit some type of action to change the object—to make it pleasing, aesthetic, or beautiful. Such is what occurs in the case of breast cancer. Breast cancer treatment typically attempts to remove the locus of the breast cancer and stop any further progression in order to save the breast. Such intervention is known as "breast-sparing" or "breast-conserving" treatment. If the breast cannot be saved, the breast is removed and breast reconstruction is offered.

Because of the psychological and physical toll experienced by patients who have a breast damaged or removed, the US Congress passed the Women's Health and Cancer Act of 1998, which mandates that all insurers cover breast reconstruction for all breast cancer patients who choose it (US Department of Labor 2013). Breast reconstruction is seen as medical treatment on the grounds that losing or damaging a breast after cancer surgery is psychologically and physically devastating and that reconstruction is a way for a woman to regain what was lost. Data show that 42% of women choose breast reconstruction following a mastectomy. The percentage goes up to about 70% as income rises (Doheny 2014). Of course, some do not choose reconstruction because they do not wish to undergo further surgeries or they do not wish to be defined by society's sense of what is beautiful (Ericksen 2008, 164; Love 2010, 47). Either way, aesthetic judgments are made and enter into how breast cancer is understood and how it serves as a treatment warrant.

Finally, breast cancer involves choices. Disease involves a choice on the part of patients at some point in history to bring a problem to the attention of health care providers. It involves a choice on the part of health care providers to recognize, describe, classify, and treat a clinical problem in a certain way. It involves a choice to fund research on particular clinical projects and rally on behalf of certain health policies and laws. It involves a choice about who makes decisions about diagnosis and treatment, what choices ought to be made and why, and who is the ethical or moral authority in such situations. In this way, disease involves *ethical* or *moral* considerations, where ethical is understood as that which is praiseworthy or blameworthy (Engelhardt 1996, 226). On this view, ethical values assign worth or significance to that which is praiseworthy or blameworthy.

In the case of breast cancer, calling a clinical condition "breast cancer" involves the choice to recognize it as such. It involves the choice on the part of some patients to seek preventive screening for breast cancer. It involves a choice on the part of some patients with certain signs and symptoms of the breast to seek medical attention. It involves a choice to stage and grade breast cancer in a way that addresses problems that arise from abnormal breast cells. It involves a choice to treat a condition in light of its diagnosis and in ways that benefit patients and their related parties. It involves a choice to fund breast cancer research and rally on behalf of policies and laws that guide breast cancer care in ways that are good for patients and related parties. It involves a choice about who makes the decisions about breast cancer diagnosis and treatment, what choices ought to be made and why, and who is the ethical or moral authority in such situations (for more on the ethical issues, see Chapter 7). In this way, our understanding of breast cancer involves choices regarding what is good and proper function, what ends are to be sought, what is aesthetically pleasing, and what is the right or good thing to do.

As seen here, disease is an evaluative notion. It is understood as an abnormality that falls within the purview of medicine to address. It represents a harm, the prevention, removal, or minimization of which is a goal of medicine. It represents an unaesthetic state and one that elicits judgments regarding what is beautiful and what is ugly. It is a choice, one made in light of clinical evidence and individual preferences. In the case of breast cancer, breast cancer involves dysfunctional cells that, if left alone, can lead to harm. It is a phenomenon in medicine that serves as a treatment warrant that is geared toward reducing harm to patients. It is unaesthetic, especially in a culture that places high value on the beauty of the breast and its role in how sexuality, women's bodies, and reproductive fitness are viewed. It is the basis of various choices, choices regarding how it is understood and choices about what actions are to take place in light of the achievement of specified clinical goals. Given the role of values in our understanding and treatment of breast cancer, one might ask once again about the extent to which the certainty of our understanding and treatment of breast cancer is called into question, a topic that is considered next.

VALUE CERTAINTY

Our discussion of the clinical values that frame breast cancer and its treatment returns us to a discussion of the certainty of our understanding of breast cancer. An earlier discussion in Chapter 3 focuses on the extent to which our understanding of breast cancer was *epistemologically* certain. It concluded that we need to revisit our assumption of cognitive certainty in breast cancer medicine and the way we navigate the uncertainty of breast cancer diagnosis, prognosis, and treatment. Given the role played by values in our understanding and treatment of breast cancer, we now explore the extent to which the values in our understanding and treatment of breast cancer make the diagnosis, prognosis, and treatment of breast cancer even less than certain.

At first, it appears that the functional, instrumental, aesthetic, and ethical values that frame our understanding and treatment of breast cancer and the actions that follow in response to this framing are universal and transcend particular communities of breast cancer specialists. Consider that breast cancer is a clinical problem worldwide and the TNM staging system of classification is used by clinicians around the world to stage and grade breast cancer (World Health Organization 2014). Given this, it appears that the diagnosis and treatment of breast cancer are based on the same set of clinical values involving minimizing patient suffering and maximizing patient welfare, objective values that are shared around the world and transcend particular communities.

In philosophy, the view that the values that frame phenomena are absolute and unchanging is known as *value objectivism.* Value objectivism is the view that universally valid or true values exist. More specifically, it holds that values (1) are objectively valid, (2) must be universalizable, (3) are interpersonal, (4) apply to the self, and (5) are knowable (Pojman and Feiser 2011, 16–17). Value objectivism is associated with the thinking of Oxford philosopher W. D. Ross (1877–1971). Ross holds that humans have an intuitive or self-evident knowledge of rightness and wrongness in terms of action-guiding principles, such as to fulfill one's promise, to refrain from harming others, and to promote justice. As he says in "What Makes Right Acts Right?": "that an act, *qua* fulfilling a promise, or *qua* effecting a just distribution of good . . . or *qua* promoting the good of others . . . is *prima facie* right, is self-evident" (Ross, in Pojman and Feiser 2011, 323). By "prima facie," Ross does not mean "an appearance which a moral situation presents at first sight" (Ross, in Pojman and Feiser 2011, 321). He rather means "an objective fact involved in the nature of the situation, or more strictly in an element of its nature" (Ross, in Pojman and Feiser 2011, 321). By "self-evident," Ross means "not in the sense that it is evident from the beginning of our lives, or as soon as we attend to the proposition for the first time, but in the sense that when we have reached sufficient mental maturity and have given sufficient attention to the proposition [,] it is evident without any need of proof, or of evidence beyond itself" (Ross, in Pojman and Feiser 2011, 323–324).

In philosophy of medicine, the view that the values that frame medical knowledge are objective is called *naturalism.* Proponents of naturalism hold that the truth of a value is independent of individual or social decisions and is based on natural properties that can be justified empirically. One recalls here the view of bioethicist and physician Edmund D. Pellegrino and bioethicist David Thomasma (1981) who define "value" in terms of naturalist criteria. As they say, value is "a property of objects capable of having action directed toward them, a property of attributes directed toward such objects, and a property of the interaction itself" (1981, 181). Consider the value of "health." For Pellegrino and Thomasma, "[h]ealth is capable of having actions directed toward it; it is a description of attitudes human beings have about their bodies; and it cements the bond of interchange between physician and patient" (1981, 181). In this way, health is a value, one found in the natural properties of being a recipient of action, a positive human attitude, and that which can be shared among humans.

Analogously, disease is a value, one found in the natural properties of being a recipient of action, a negative human attitude, and that which can be shared among humans. In the case of breast cancer, breast cancer is considered a treatment warrant in medicine. It is the focus of negative attitudes expressed by patients, related parties, clinicians, and researchers. It is a condition marked by dysfunctional cells that undermine human good and that can lead to unaesthetic states of affairs. Breast cancer is bad and, as such, ought to be treated. In this way, breast cancer is the focus of the clinician–patient relationship, one that is geared toward curing, if not preventing or managing, the clinical condition. On this view, breast cancer is an objective value located in natural properties that can be justified empirically and serves as a warrant for action.

Yet, despite our first impression, the values that frame our understanding of breast cancer may not be as objective as we may first think. Consider the challenge in establishing objectivity in interpretations of function, means to ends, beauty, and choice in our understanding and treatment of breast cancer. Regarding function, some hold that hyperplasia (i.e., overgrowth of the cells) in breast ducts should be treated, given that we do not know which cases of hyperplasia develop into late-stage cancer and which do not (Beck 2012). Others hold that hyperplasia should *not* be treated, at least not aggressively, because to do so is to overdiagnose and overtreat a condition that is not breast cancer (Lerner 2001, 246–248). Regarding means to an end, some hold that certain types of interventions (e.g., a mastectomy) are better than others (e.g., a lumpectomy) in treating late-stage breast cancer (see Figure 4.2; National Cancer Institute 2014a). Others hold that breast cancer specialists need to move beyond less-than-specific and debilitating forms of surgical treatment for breast cancer and find more targeted therapies for breast cancer. Regarding what constitutes beauty, some praise the efforts of breast cancer advocates who continue to lobby for coverage for reconstructive surgeries for breast cancer patients. Others hold that the absence of a breast should not be considered unaesthetic and women should not feel that they must go through harsh reconstructive surgeries that can lead to all sorts of complications in order to be considered beautiful or whole again (Erickson 2008). Regarding choice, some hold that certain choices are considered praiseworthy while others are blameworthy, thus leading to an array of differences in views about what ought to be done in treating breast cancer. Not only is the role of patient choice accepted in breast cancer medicine, it is encouraged, as in the case when clinicians turn to newly diagnosed breast cancer patients and ask them to make decisions about what approach to treatment they prefer. The point is that the values that frame breast cancer appear to be anything but objective.

The philosophical view that values are not objective is called *value subjectivism.* Value subjectivism holds that values are an expression of personal or private preference. What is of value for one person may differ radically from what is of value for another (Pojman and Fieser 2011, 17). Value subjectivism is associated with the thinking of the early Sophist Protagoras of Abdera (481–411 B.C.E.), who is reported by philosopher Plato in *Theaetetus* to say that "each of us is the measure of the things that are and those that are not; but each person differs immeasurably from every other" (in Reale 1987, 166D, 160). The view expressed by Protagoras is that what is determined to be of value is dependent upon the individual. Put another way, the individual is the source and standard of what is considered valuable or of worth.

In philosophy of medicine, it is difficult to find advocates of value subjectivism. This is because medicine as an institution is typically committed to some objective version of advancing the best interest of patients and minimizing the harms brought about by disease, illness, and deformity. Nevertheless, there are voices critical of the view that we can have objective access to values in medicine, although they typically stop short of full-blown value subjectivism. For instance, and addressing whether there are objective values in bioethics, Engelhardt claims that "[W]estern morality and the moral authority of public policy have been grounded in the assumption that reason could justify a canonical content-full morality and bioethics for moral strangers" (1996, 83). Engelhardt finds this assumption problematic because reason is unable to deliver objective knowledge of values. It cannot because of the limits of reason. Sound rational argument is unable to deliver particular concrete value viewpoints in any canonically decisive way. It cannot because it does not have the ability to do so. Nevertheless, there is hope. As Engelhardt says, "although reason fails to convey moral authority, a moral authority binding moral strangers can be derived from the permission of those who choose to collaborate" (1996, 83). In this way, "[a] limit can be given to the nihilism [i.e., the rejection of the possibility of knowledge] that challenges us: general cannons of secular morality can indeed be articulated" (1996, 83).

For Engelhardt, agreement or permission grounds the canons of a secular value framework. Agreement provides a moral basis for authorizing certain actions over others. According to Engelhardt, there will be circumstances "that will likely be valued across communities" (Engelhardt 1996, 204). Applied to breast cancer, the pain and suffering associated with breast cancer will be considered, for the most part, a disvalue. What follows from this judgment of disvalue is the sense that breast cancer is bad and ought to be treated. Alternatively, there will be circumstances that will not command this type of "unifying assent" (Engelhardt 1996, 204). Applied to breast cancer, judgments regarding the assessment of risk associated with the different stages of breast cancer turn not only on current clinical evidence but personal views regarding what one is willing to live with, undergo, and gamble in light of clinical evidence—and sometimes in spite of it. In addition, assessments of breast beauty express a wide range of evaluations among individuals and across communities, thus leading to varying preferences toward treatment. In the end, value uncertainty pervades clinical knowledge, but not at the cost of taking values seriously in medicine, a topic that is next considered.

VALUE SKEPTICISM

We find ourselves once again returning to the discussion in Chapter 3 about skepticism. We learned in Chapter 3 that there is a variety of skeptical responses to the claims that frame breast cancer, ranging from absolute to philosophical to common sense. This lesson applies as well to skeptical responses to the values that frame breast cancer. Absolute skeptics will have little interest in the values that frame breast cancer because they fundamentally distrust medicine and the values that play a role. Here we can say that an absolute skeptic will likely not enter into the medical arena for breast cancer screening or testing. Philosophical skeptics will be inquisitive about the values that frame breast cancer and be willing to question even the most cherished evaluative claims in breast cancer medicine. An example here might be a patient who questions the current way breast cancer is diagnosed and treated and entertains alternative ways breast cancer can be understood and treated. Common-sense skeptics will also be inquisitive and ask numerous questions concerning breast cancer. An example here is a patient who accepts clinicians' recommendations and questions her health care providers about the diagnosis and her care. It follows once again that a healthy sense of skepticism drawn from philosophical and common-sense skepticism will be helpful in navigating the evaluations that are made in breast cancer care. Certain values such as the minimization of pain and suffering and the thwarting of death can be expected to be widely shared in breast cancer medicine. Other values such as quality-of-life determinations and reactions to risk assessments can be expected to differ among patients and breast cancer specialists. In the end, while there may not be clear and unambiguous values that guide all decisions in breast cancer medicine, it does appear that nihilism is not an inevitable option. General senses of values can and will be articulated in our understanding and treatment of breast cancer.

Drawing on an earlier discussion of Foucault (see Chapter 2), there is a sense of certainty in evaluation that arises in relation to the number of evaluations that are made and expressed. Value certainty arises from a multiplicity of individual evaluations that are expressed and shared. This does not mean that an individual cannot herself value something individually and in most important ways. Rather, the values that guide us in collective endeavors such as breast cancer care will arise from a multiplicity of sources and expressions. Such multiplicity becomes a vehicle of an index of convergence and divergence. In breast cancer medicine, the values of maximizing patient welfare and minimizing patient harms will guide evaluations in breast cancer medicine because they are widely supported by those participating in breast cancer medicine. In this way, there is evaluative unity in diversity in that there will be widely shared values among clinicians and patients about breast cancer diagnostic, prognostic, and therapeutic criteria. And there will be diversity in unity in that there will be legitimate differences in how clinicians and patients evaluate their options. Ultimately, they evaluate their options in light of select principles and goals.

CLOSING

Philosophically speaking, the question about how breast cancer is evaluative raises a host of considerations, including normativism, the kinds of clinical values in medical thinking, value objectivism, and value skepticism. What we find is that, initially, breast cancer is a treatment warrant and appears to fit the view of a clinical entity that is value-neutral. But things are not as simple as one would initially think. Upon reflection, descriptions and explanations of breast cancer are nested in evaluative frames of reference through which they are seen, interpreted, and responded to. Clinical evaluations of breast cancer are complex and involve appeals to functional, instrumental, aesthetic, and ethical values. Such values are not as objective as we may think at first. As a consequence, clinicians and patients alike must face the recognition that clinical evaluations of breast cancer are to some extent uncertain and a healthy sense of skepticism provides a check against an idealized sense of evaluation in breast cancer medicine. This does not mean that anything goes, for the central and shared clinical values of addressing dysfunction, minimizing patient harm and maximizing patient welfare, maintaining beauty and symmetry of the body, and honoring choice in light of limited clinical evidence and individual preferences frame our understanding and treatment of breast cancer.

Chapter 5

How Is Breast Cancer a Social Phenomenon?

PERSONAL MUSINGS

As a breast cancer patient, I often reflected upon my new identity as a breast cancer patient. I have been called a "survivor" and "warrior" and find myself questioning the status of both designations. I do not see myself as a survivor because I am not sure what I have survived. I know that I have lived through all my surgeries, but I do not think that is the main intent of the label. (And if it is, what does this say about the current breast cancer treatments?) Although I have completed my initial breast cancer treatments, I know that I cannot be sure if I will ever be "cured." In this sense my survival was and continues to be questionable, and hence I can never be rest assured about it. I also do not see myself as a warrior as are my military students in Colorado Springs, Colorado, who have come back home from the wars in Afghanistan and Iraq. Those who have bravely served us in battle are warriors, not someone like me who has been all too willing to whine and give up the breast cancer fight at a moment's notice. I also do not see myself as a warrior because I have had many moments of simply not wanting to know what is to come. For instance, I am often thankful that I had no sense of what I was getting into when I agreed to a double mastectomy and reconstructive surgery. Thank goodness, prior to my mastectomy and direct reconstruction, I did not consider the extent to which I would be cut, filled, and stitched. Thank goodness I did not consider how I would feel post-surgery. Given this, I'm no warrior, despite what the October media blitz suggests.

So the question arises, who am I now that I have breast cancer? Who am I now that I have lost my breasts, undergone numerous surgeries and their aftermaths, and become aware that a terrifying disease has struck and can return at any time? Amidst this confusion of identity, I can be sure of one thing: that I am not alone in my journey as a breast cancer patient. I have been surrounded by family and friends, and have met so many others who have gone down my path. I have become a part of a social club that I did not ask to join, and members of the club have been more than generous with their time and services. Such musings have led me to reflect on the social character of breast cancer. To what extent is the diagnosis and treatment of my breast cancer framed by collaborative forces? How do these forces create certain perceptions and experiences of the breast cancer patient in general, and of me in particular? Given that collaborative forces are at play in our understanding of breast cancer, to what extent is breast cancer medicalized or constructed? Given the role played by public policy and law in breast cancer care, to what extent is our understanding and treatment of breast cancer a result of the politicalization of medical reality? Herein lay the basis of some of my philosophical reflections on the social dimension of breast cancer and the sociological issues concerning the collaboration, medicalization, commodification, and politicalization of clinical reality. Here, and in the context of socializing breast cancer, acollaboration is contrasted with collaboration, amedicalization is contrasted with medicalization, acommodification with commodification, and apoliticalization is contrasted with politicalization (Figure 5.1).

(a)	acollaboration (the view that knowledge is not communal)	collaboration (the view that knowledge is communal)
(b)	amedicalization (the view that a non-medical phenomenon is not a medical one)	medicalization (the view that a non-medical phenomenon is a medical one)
(c)	acommodification (the view that a non-economic phenomenon is not an economic one)	commodification (the view that a non-economic phenomenon is an economic one)
(d)	apoliticalization (the view that a non-political phenomenon is not a political one)	politicalization (the view that a non-political phenomenon is a political one)

Figure 5.1 Sociological issues.

COLLABORATION

There is a tendency in medicine to view breast cancer as the finding of isolated individuals who have made remarkable discoveries. This is in part because we tend to organize history around individual discoveries. In the case of breast cancer, consider Egyptian Imhotep's description (approx. 2500 B.C.E.) of a "bulging mass in the breast" (Mukherjee 2010, 41), Greek historian Herodotus' account (approx. 450 B.C.E.) of inflammatory breast cancer (Mukherjee 2010, 41), Greek physician Hippocrates' account (approx. fifth century B.C.E.) of breast cancer (or what he called *karkinos*) in terms of "a crab dug in the sand with its legs spread in a circle" (Mukherjee 2010, 47), German pathologist Rudolf Virchow's account (1858) of cancer as "*omnis cellula e cellula*" (meaning "where a cell arises, there a cell must have previously existed") (Mukherjee 2010, 15), and University of

Utah geneticist Mary-Claire King's account (approx. 1990s) of breast cancer as a genetic condition (Olson 2002, 254–255). The initial impression is that our understanding of breast cancer is collaboratively neutral.

In philosophy, the view that knowledge is collaboratively neutral is what I call an *acollaborative* view of knowledge. An acollaborative view of knowledge understands isolated individuals to be the source of knowledge. This asociological view of knowledge is found in the thinking of Greek philosopher and mathematician Pythagoras (approx. 600 B.C.E.) who likens the philosopher or knower to a spectator at the ancient games. As he says, "when Leon the tyrant of Philius asked him who he was, he said, 'A philosopher,' and he compared life to the Great Games, where some went to compete for the prize and others went with wares to sell, but the best as spectators" (in Laertius 1925, VIII, 8). For Pythagoras, the philosopher as an individual spectator is the epitome of the knower: "in life, some grow up with servile natures, greedy for fame and gain, but the philosopher seeks truth" (in Laertius, 1925, VIII, 8). Here the knower of truth is portrayed as an individual who comes to know reality separate and distinct from others.

The image here is that the knower is a solitary individual discovering a house of truth upon a foundation of clear and distinct ideas. As feminist philosopher Lorraine Code puts it, the knower is depicted as a "featureless abstraction" (Code 2003 [1991], 559). By this, she means that "she or he [as a knower] is merely a placeholder in the proposition 'S knows that p" (Code 2003 [1991], 559). In this understanding of knowledge, "the central 'problem of knowledge' has been to determine necessary and sufficient conditions for the possibility of justification of knowledge claims" (Code 2003 [1991], 559–560). In this endeavor, philosophers seek "ways of establishing a relation of correspondence between knowledge and 'reality' and/ or ways of establishing the coherence of particular knowledge claims within systems of already-established truths" (Code 2003 [1991], 560). They tend to think that the question "Who is S?" is neither legitimate nor relevant. On this view, knowledge is typically seen as value-neutral and acontextual.

In philosophy of medicine, there is an enduring image of the clinician and the patient as an individual knower working in isolation from the broader culture in which she or he lives. Medical books on clinical decision making are typically written in terms of what a clinician knows, should know, or should ask about. Those on informed consent are typically written in terms of the image of the patient as an independent, rational decision maker. Chapters on the physician–patient, or research subject–clinical investigator, relationship typically feature a one-on-one relationship without much attention to the social context in which the relationship operates and how clinical decision making takes place within such contexts. In breast cancer medicine, the portrayal tends to be a clinical specialist meeting with an individual patient to discuss the results of clinical tests and the options that are available to the patient. The patient is seen to be an autonomous decision maker who makes choices in light of the clinical evidence as well as her desires, wishes, and goals. The image is that medical knowledge is acollaborative and an expression of logical internal criteria and an epistemological methodology that do not depend on contexts.

But upon further consideration, our understanding of breast cancer is not collaboratively neutral. As University of Pennsylvania physician and historian of medicine Robert Aronowitz puts it, "Social norms and attitudes, not only clinical and technological developments, have determined how we classify and diagnose cancer" (2007, 10). Such norms and attitudes handed down in time influence our understanding of cancer. They steer us to hold particular views of breast cancer and expect certain actions with a diagnosis of breast cancer. They come about through the contributions of researchers and clinicians working together in research labs at institutes, companies, and university hospitals. One thinks here of the Mayo Clinic, Memorial Sloan Kettering, and Cancer Treatment Centers of America as well as the Department of Breast Surgery or Breast Oncology at the University of Colorado. In addition, such norms and attitudes come about through the contributions of members of the lay public who devote great time and energy toward such causes. One thinks here of the great contributions of Shirley Temple Black, Rose Kushner (1975), Betty Rollin (2000 [1976]), Betty Ford, Happy Rockefeller, Audre Lorde (1980), Nancy Reagan, and Barbara Ehrenreich (2001) (Lerner 2006). Collaborative forces frame our understanding of breast cancer.

The collaborative forces that frame disease can be formal or informal (Engelhardt 1996, 219). Breast cancer specialists themselves operate within collaborative environments in their practices, their specialties, their educational environments, and the clinics and hospitals in which they practice. Some of the formal collaborative endeavors include developing professional clinical standards, devising educational requirements and licensure agreements, formulating funding options, and instituting health laws and policies. In addition, formal practices, such as meetings of the American Joint Committee on Cancer (2010), organize rules for the use of descriptive terms in diagnostic categories for cancers. In such circumstances, the collaborative reality of disease is settled by votes within committees. "The decisions in such circumstances are made not simply in terms of the character of reality as it is taken really to be, but also in terms of which modes of classification will be most useful in organizing treatment and care" (Engelhardt 1996, 219). Choices to divide breast cancer stages and grades into a certain number turn on cost–benefit calculations and understandings of prudent actions that have direct implications for the ways patients are treated (Cutter 1992). In addition, formal practices allow for the possibility of organizing large-scale campaigns for purposes of raising funds for breast cancer practice and research as well as the distribution of resources to breast cancer patients.

Informal collaborative practices also frame our understanding of disease. Breast cancer researchers and practitioners themselves operate within informal collaborative environments in which they make decisions about what they wish to focus on in their practices, what groups of clinical professionals to join, where they wish to locate, what patient populations they wish to see, and what they are willing to be paid for their work. Patients and research subjects themselves operate within informal collaborative environments in which they blog about their health care providers, clinical conditions, diagnoses, treatments, and recommendations, and advocate on behalf of increased attention to particular clinical problems. Their comments and actions have influence on what diseases are given attention and which ones are not (Lerner 2001). Clinical researchers and practitioners, and subjects and patients, contribute to the informal social networks that frame our understanding of breast cancer.

The collaborative practices that frame our understanding of disease in general, and of breast cancer in particular, are not incidental. They are central to knowledge construction, acquisition, and application. Humans as knowers operate within "material, historical, [and] cultural circumstances" (Code 2003 [1991], 562) through which they come to know. Such a "perspective" or "standpoint" is collaborative and set within particular local contexts. Such a *collaborative view* of knowledge focuses on how the social locations of knowers affect what and how they know (Anderson 2011). The focus is not on criteria of evidence, justification, and warrantability, but rather on the identities of cognitive agents. The emphasis is on the character, local circumstances, and interest of a knower in an inquiry. According to Code, this emphasis raises "questions about how credibility is established, about connections between knowledge and power, about the place of knowledge in ethical and aesthetic judgments, and about political agendas and the responsibilities of knowers" (Code 2003 [1991], 562–563).

A collaborative view of knowledge in contemporary philosophy of science is rooted in the thinking of University of Chicago philosopher of science Thomas Kuhn (1922–1996). Kuhn argues that the history of science is not a smooth progressive accumulation of data and successful theory, but the outcome of ruptures, false starts, and constraints. Science works within frameworks of assumptions that Kuhn calls "paradigms," which change and shift depending on many different variables, including social ones. In science, and by implication medicine, a paradigm is not an object of replication. As Kuhn says, "[i]nstead, like an accepted judicial decision in the common law, it is an object for further articulation and specification under new or more stringent conditions" (1970, 23). In this way, paradigms function heuristically to guide our thinking; they help solve problems that a group of practitioners deems to be acute (Kuhn 1970, 23). In this way, paradigms draw on communal knowledge and practice.

In philosophy of medicine, a collaborative view of knowledge is found in the thinking of Polish-Israeli physician and philosopher of medicine Ludwik Fleck (1896–1961), whose work Kuhn admired. Fleck argues that there are no such things as neutral, naked, or bare facts. As he says, "we can define a scientific fact as a thought-stylized conceptual relation which can be investigated from the point of view of history and from that of psychology" (1979, 83). Facts always appear interpreted within the embrace of theoretical frameworks, whether or not these frameworks are formally or informally developed. In addition, they are interpreted within sociohistorical frameworks, whether or not these frameworks are formally or informally developed. The expectation of a timeless or transcendent account of medical reality is not forthcoming because we as knowers operate within sociohistorical contexts that frame our reality, knowledge, and values.

In his work, Fleck reconstructs the development of the modern concept of syphilis from the early modern concept of "carnal scourge" to the midmodern "empirical-pathological" concept (treatable with the nonspecific agent mercury) to the late modern "empirical-pathological concept" (treatable with the specific antibacterial agents arsphenamine and neoarasphenamine). In the course of time, the concept of syphilis changes "from the mythical, through the empirical and generally pathogenetical, to the mainly etiological" (1979 [1935], 19). A major lesson here is that the modernist expectation of a timeless, acollaborative account of medical reality is not forthcoming because humans live in changing environments, which change them as well.

The Fleckian-Kuhnian view of the knower as a social being who operates within a particular context has been developed and revised further by contemporary feminist sociologists and philosophers. A number of accounts have been advanced that address different social models of knowers, knowledge and objectivity, and epistemic values (Grasswisk 2013). Where there are significant differences among accounts, there are shared points of discussion on "the effects of social relations internal to epistemic communities, social relations external to those communities, and social relations between knowledge-producing communities and lay communities" (Grasswick 2013, 22). On this view, what counts as a knower and as evidence is framed by a community's theories, value commitments, and observations. There is neither a knower nor knowledge apart from such a shared complex. The community is the primary knower and individual knowledge is dependent on the knowledge and values of the community (Nelson 1990; also see Fuller 1988).

In the case of breast cancer, a timeless, acollaborative account of breast cancer is not forthcoming. As New York University physician and historian of medicine Barron Lerner says, "The history of breast cancer in the twentieth century is replete with examples of seemingly objective terminology, the actual meanings of which reflect the cultural context into which they were introduced" (2001, 11). Think about it: our understanding of breast cancer has changed from the Ancient concept of "bad bile" to the modern "empirical-pathological" concept of a "breast tumor" (treatable by surgical techniques) to the early contemporary "experimental-pathological concept" of "dysfunctional cells" (treatable by surgical removal of the tumor, chemotherapeutic agents, and/or radiation) to the twenty-first century "biomolecular-pathological concept" of "dysfunctional genes" (to be treated by genetic interventions). In the course of time, the concept of breast cancer changes from a mythical through an anatomical and physiological to a biomolecular account. Here, the expectation of an acollaborative account of breast cancer is not forthcoming because our understanding is embedded within contextual frameworks that change and evolve.

On this view, our understanding of breast cancer reflects the cultural context in which it is introduced. It reflects the social and power relations internal to the knowledge-producing community of breast cancer specialists and relates current information about breast cancer in light of accepted clinical standards and practices. An example here is the influence held by professional organizations, such as the American Society of Breast Surgeons and the American Society of Clinical Oncology, each of which advance positions on various topics in breast cancer medicine. In addition, our understanding of breast cancer reflects the social and power relations external to those communities, such as those found in advocacy groups, such as the Susan G. Komen and Avon Foundation, in charting directions for breast cancer care and research. It reflects the social and power relations between the knowledge-producing communities and the lay communities, and in particular those that foster a dialogue between them for purposes of advancing breast cancer care. An example here is the thinking that arises out of conferences, such as the San Antonio Breast Symposium, in which both knowledge producers and lay individuals together discuss matters of importance in breast cancer medicine. Given the social context of breast cancer, issues arise concerning what constitutes a legitimate focus of breast cancer medicine, a topic that is next considered.

MEDICALIZATION

As indicated in Chapters 2 and 3, there is a tendency in medicine to view breast cancer as a finding based strictly on biomedical evidence. On this view, breast cancer is understood in terms of deviations from the norms of measurable biological variables and explained on the basis of dysfunctional biological processes. As an example, the TNM staging system (see Chapter 3) organizes different biological events of the breast in terms of their deviation from the norms of cellular structure and function. The designation of the T, N, and M levels of the tumor indicates the size, extent, and spread of the cancer, respectively. The TNM system is then used to guide therapeutic interventions in medicine, which are submitted to clinical scrutiny to assess their success rates. Accordingly, it is revised in light of new knowledge about breast cancer and the need to maximize patient benefit and minimize patient harm.

In philosophy, the view that there is an objective approach to studying disease is called *amedicalization* or *scientific thinking*. As Gøtzsche describes it, "'Scientific thinking' is characterized by a critical attitude to established theories on the part of the scientists. They seek weak points in the theories, generate new hypotheses and subject these to critical experiments" (2007, 112). "Scientific thinking" is rooted in the thinking of the early modern scientists such as

Galileo Galilei (1564–1642), René Descartes (1596–1650), and Isaac Newton (1642–1727). Galileo, Descartes, and Newton were all interested in developing a methodology for knowing the natural world based on careful and rigorous observation. Scientific thinking is analytical, "meaning that entities to be investigated be resolved into isolatable causal chains or units from which it was assumed that the whole could be understood, both materially and conceptually, by reconstituting the parts" (Engel 1981 [1979], 593–594). Scientific thinking is rational, in that it makes logical sense; it is empirical, in that it is based on what is observed. In this framework, the body is understood as a "machine," disease as a consequence of the breakdown of the machine, and the health care professional's role as the technician of the machine (Engel 1981 [1979], 594).

In philosophy of medicine, "scientific thinking" is called *medical thinking*. Medical thinking about disease occurs when certain criteria are satisfied. According to Georgetown University biomedical ethicist Robert Veatch, medical thinking about disease takes place when the condition is seen as (1) "non-voluntary," (2) "organic," and (3) "falls below some socially defined minimal standard of acceptability," as well as (4) when the expert is "the class of relevant, technically-competent" physicians (Veatch 1981, 530). As an example, medical thinking about breast cancer takes place when the condition is seen to be something that occurs through no fault of the patient. The condition is seen to be organic and found in nature, or the biology of the body, and falls below a socially defined minimal standard of acceptability. It is deemed a clinical problem by health care professionals not only because it represents biological dysfunction of the breast, but because it can lead to pain and suffering and, in some cases, the death of the patient.

But come to think about it, things are not so simple in the case of breast cancer. *Nonmedical thinking* pervades medical thinking. The social forces that frame clinical reality can bring about what is called in the literature the *medicalization* of clinical problems. Medical sociologist Peter Conrad (2007) defines medicalization as "a process by which nonmedical problems become defined and treated as medical problems" (2007, 4). In this process, nonmedical problems are "defined in medical terms, described using medical language, understood through the adoption of a medical framework, or 'treated' with a medical intervention" (Conrad 2007, 5; also see Conrad and Schneider 1980). In a lengthy analysis, Conrad illustrates how, over the past half-century, the terrain of health and illness has been transformed. What were once considered normal human events and common human problems, such as birth, menstruation, pregnancy, menopause, and aging, are transformed into medical conditions and treated as such. They are intentionally or unintentionally transformed into medical problems because medicine offers interventions that can change their occurrences and patients seek medical attention for conditions that are unwanted, even though they may not be considered traditional medical problems. Conrad (2007) warns that the medicalization of illness, dysfunction, and disorder has far-reaching implications in our lives. It places labels on clinical events that might better be understood as part of "normal" life and instructs that such "normal" events should be treated with medical interventions. This is a problem because medical interventions carry their own set of troublesome short- and long-term consequences that are better off avoided, if at all possible.

At first, it sounds odd, and perhaps downright irresponsible, to suggest that we medicalize breast cancer. It sounds odd because I really had (and perhaps still have) breast cancer. It sounds downright irresponsible because breast cancer really occurs and annually takes the lives of approximately 40,000 women and 400 men in the United States (National Cancer Institute 2014a, 1; "Breast Cancer in Men" 2014) and approximately 450,000 individuals worldwide (World Health Organization 2013, 1). But bear with me for a moment. Aronowitz (2007) shows in a lengthy analysis how clinicians, public health officials, and American society have transformed the understanding and experience of breast cancer through overselling the effectiveness of breast cancer screening and treatment (2007, 2). They have oversold medical screening and treatment by advocating for widespread screening and treatment for breast cancer despite not having clear data about which precancerous and early-stage breast cancers develop into late-stage breast cancers and which do not. They have advocated for widespread screening and treatment despite the risks brought about by the clinical interventions.

One consequence of having less-than-certain screens, tests, and treatments for breast cancer is the medicalization of conditions as breast cancer that are not breast cancer. Consider, for instance, ductal carcinoma in situ. Ductal carcinoma in situ is considered in the TNM staging system "Stage 0 carcinoma." It is considered by some clinicians as precancer and others as cancer. Why this is the case in medicine is because of medicine's inability to predict which early-stage cancers develop into late-stage cancers and which do not, so it is prudent for clinicians to err on the side of treating (Beck 2014). As Aronowitz summarizes the problem, "The juxtaposition of 'carcinoma' with 'in situ' to form a name for an uncertain probability of future cancer is part of a larger phenomenon in which the terms used to denote, modify, or describe cancer and cancer risks carry problematic connotations, which influence our clinical and policy responses" (2007, 11). One of the problematic connotations is diagnosing a condition as breast cancer that is not breast cancer. Estimates are that there are up to 50% false-positive diagnoses of women screened for breast cancer (Boyles 2012). Another is justifying medical interventions, such as tests, biopsies, lumpectomies, mastectomies, and breast reconstructions, that may not be needed and may carry serious risks for patients.

Another example of the medicalization of breast cancer can be taken from the breast reconstruction literature. Medical sociologist Nora Jacobsen (1998) illustrates how the introduction of breast implant techniques has shaped how women see breasts in an idealized fashion and has led to what she calls "the medical construction of need." The view here is that women come to believe that they need breast reconstruction after having their breast(s) removed by a mastectomy. They come to believe this because of societal expectations of what a woman's breasts, as well as a woman, should look like, replete with "man-made" cleavage and curves (Jacobsen 2000). The reconstructed breast becomes a symbol of the patient's return to sexual and reproductive fitness. Further, with breast reconstruction often come additional breast interventions, such as multiple breast revision surgeries that address scarring, capsular contracture, and rippling (US Food and Drug Administration 2015). Still further, given that implants need to be replaced every ten to twenty years, even further surgeries will be required. All of this sets the stage for what some call the medicalization of particular kinds of breast conditions.

Seen here, the medicalization of breast cancer raises numerous concerns. There are concerns about medicine defining nonmedical events as medical in a "camoflagued" way and under the guise of medical authority. As social scientist Irving Kenneth Zola says, "[m]edicine is becoming a major institution of social control" (1972, 487). It has become this "by making medicine and the labels 'healthy' and 'ill' *relevant* to an ever increasing part of human existence" (1972, 487). The motivation to accomplish this is rooted in the search for authority, the desire to make a profit, and the quest to empower a profession that has grown in power during the last century. It is only in the late nineteenth century when medicine began to achieve professional status that commanded a level of respect from the public. With this came the initial efforts of medical professionals to make a profit from their work. These efforts come together in the twentieth century in medicine's quest to become a self-regulating, economically sustaining profession that offers expertise on matters concerning disease, illness, and health (Starr 1982).

And yet, it is not only members of the medical profession who call for medical intervention for nonmedical matters. Patients do as well. According to Aronowitz (2007), patients contribute to the medicalization of phenomena by requesting or demanding tests or interventions for things that concern them. Some are motivated by being overly cautious and some by fear and anxiety brought on by aggressive screening and testing campaigns. As Aronowitz paints the picture, "there has been a war against time fought on a series of fronts—public education campaigns to get women not to delay in seeking medical attention for lumps and other suspicious breast changes, technological innovation aimed at developing new ways of diagnosis, new pathologic knowledge about 'earlier' stages in the natural history of breast cancer, and . . . aggressive treatments aimed against the possibility of future disease" (Aronowitz 2007, 15). All of this adds up to a number of forces convincing women that they need to act and act fast in the face of some breast aberration. It is as if the question is "When will I get breast cancer?" as opposed to "What is the chance that I will get breast cancer?" As Aronowitz says, "overlapping factors are at work in shaping modern fears of breast cancer" (2007, 262). Because breast cancer is an empirical phenomenon nested within frames of evaluations and social contexts, it is replete with "nonmedical" influences. One of these influences is the commodification of the breast, a topic that is explored next.

COMMODIFICATION

At first, we tend to think that clinical phenomena are not considered a commodity, or good or service that is exchanged on the market.

Clinical phenomena are not considered a commodity because, for the most part, they occur through no fault of the patient and it would be inappropriate to support a system in which only those who can afford care can obtain it. Such is a major reason that many countries have adopted a national health care program through which citizens can seek clinical care. And even when countries do not have a national health care program, such as the United States, they offer other governmental-sponsored programs, such as Medicaid and Medicare, that provide access to health care for those who have limited resources.

In addition, clinical phenomena are not considered a commodity because they are experienced by patients, and patients need to be protected against unwarranted intrusions, such as market forces, when seeking and obtaining clinical care. Patients need to be protected because, if they are not, they are vulnerable to market forces and may not receive proper clinical care. They are vulnerable because they may not be protected from coercive forces in making choices about their medical care. If they are not protected from coercive forces, they are less able to make autonomous choices about their care in the health care setting. If they do not make autonomous choices, they may not receive proper clinical care. Society may be harmed as well because it is society that may have to take care of patients who do not receive proper clinical care. In this way, clinical phenomena, including breast cancer, are seen not to be a commodity in the market.

In philosophy, the view that clinical events are not considered a commodity is what I call the *acommodification of disease and illness.* Such a view is rooted in the thinking of Canadian physician and cofounder of Johns Hopkins hospital William Osler (1849–1919), who says: "The practice of medicine is an art, not a trade; a calling, not a business" (Cheung 2017, 70). Osler believed that medical rounds, bedside observations, close physical examination, and attention to patient complaints distinguished medicine from other lines of work, such as found in business (Bliss 1999). He taught that medicine is a science and art, and the use of technology must not undermine the patient–physician relationship and the care of patients. While Osler had no objection to having physicians paid for their work, he believed that individuals should pursue the medical profession not for money but for the love of humanity and caring for patients.

In philosophy of medicine, the issue of commodification receives wide attention in the case of cosmetic surgery. According to the American Board of Cosmetic Surgery, cosmetic surgery is "focused on enhancing a patient's appearance" (2015) using medical technology. The most popular cosmetic surgeries in the United States are breast augmentation and facial surgeries and treatments (American Board of Cosmetic Surgery 2015). According to feminist philosopher Kathryn Pauly Morgan (1991), cosmetic surgery is problematic because it takes something that should not be a commodity (e.g., breasts, women's images) and makes it into something that can be traded and sold in the market. A market for women's body parts is a problem in three ways. First is the problem of conformity: "women do not use the medical technology to underscore their uniqueness or eccentricity, rather they all let the same 'Baywatch' standard determine their looks" (Morgan 1991, 36). Second is the problem of colonization: it looks as if women are cultivating their own bodies, whereas in fact their bodies are being colonized by physicians, the majority of whom are male. Third is the problem of coerced voluntariness and the technological beauty imperative: women are coerced into thinking they need cosmetic surgery through the barrage of media and cosmetic surgery campaigns, which results in "the 'ordinary'" coming "to be perceived and evaluated as the 'ugly'" (Morgan 1991, 41). In this way, women are coerced into choosing cosmetic surgeries, and coercion is problematic because it undermines patient autonomy.

One might say, of course, that there are important differences between breast cancer surgery and cosmetic surgery. Breast cancer surgery is typically not a choice, is recommended in order to prevent the development of future harms, and is covered by all insurers in the United States. Nevertheless, breast cancer surgery is in part cosmetic surgery in that surgeons seek to reconstruct the breast following the removal of cancer sites in an aesthetically pleasing way. In this way, breast cancer surgery is vulnerable to the criticisms that plague cosmetic surgery.

There appears to be great support for the view that breast cancer and aspects of breast cancer (e.g., its treatments) ought not to be commodities in the market. They ought not to be commodities because, to consider them so would be to situate them in contexts in which they are treated as sources of revenue and profit. If treated as sources of revenue and profit, breast cancer patients may not receive the proper care for their conditions and patients as consumers could be coerced into choosing forms of treatment that are not appropriate for their conditions. In addition, to treat breast cancer and aspects of breast cancer as commodities risks the colonization of women's clinical conditions for purposes of making money or cultivating women's bodies according to societal expectations. In this way, the commodification of breast cancer is seen to be problematic not only among breast cancer patients and health care professionals but among legislators who pass laws providing breast cancer patients access to breast cancer treatment in an affordable way. Even though breast cancer patients incur costs from their breast cancer treatment, the costs are significantly reduced when limits are placed on what breast reconstructive surgeons can charge for their services.

But upon further consideration, breast cancer and aspects of breast cancer are in part commodities that can be transformed into goods and services. I will call this view the *commodification of disease and illness*. One thinks here of the market for screens, tests, and treatments for breast cancer patients. Whether they are insurers or producers of medical equipment, companies line up to create and sell the latest and greatest for doctors and hospitals to use in breast cancer medicine. In this way, breast cancer is a major market force in the medico-industrial complex. In addition, one also thinks of breast reconstruction, which involves economic incentives to sell breast implants to patients who have lost them to cancer. Still further, one thinks of "the cornucopia of pink-themed breast cancer products" (Ehrenreich 2001) that are featured in hospitals that serve breast cancer patients as well as in media campaigns, reminding breast cancer patients that they are part of a large circle of survivors, warriors, or comrades. Note that the annual Breast Cancer Awareness Month held in October is funded by Astra-Zeneca, the company that produces tamoxifen, the world's largest selling hormonal treatment for breast cancer. Not only do many breast cancer patients take tamoxifen post-cancer diagnosis, there is growing interest in recommending tamoxifen as a preventive means to avoid breast cancer, thus increasing the number of women on tamoxifen and the economic benefits to a for-profit company (DeGregorio and Wiebe 1996).

In philosophy of medicine, at least in some traditions, the sale of clinical goods and services is not necessarily a problem. In the case of cosmetic surgery, feminist sociologist Kathy Davis offers a view of cosmetic surgery that contrasts with the view advanced by Morgan summarized earlier. Davis investigates the actual decision-making process of women undergoing breast augmentation. What she finds is that such women choose cosmetic surgeries in order to address some bodily concern, put an end to their sufferings, and become more "ordinary" (Davis 1993, 164). Davis finds, in contrast to Morgan, that women carefully think through their options and decide in favor of an option when it provides the means to achieve what they seek. As she says, women were "no more duped by the feminine beauty-system than women who do not see cosmetic surgery as a remedy to their problems with their appearance" (1995, 163). More current studies show that women are for the most part satisfied with their breast cancer reconstruction and are pleased that they chose to reconstruct (Lantz et al. 2005).

In the case of breast cancer, while there are market forces encouraging women to choose one provider over another, one hospital over another, and one surgery over another, it does not follow that women are duped into choosing one option over the other. As we have learned in the previous chapters, choices will always be constrained by circumstances. As long as one has relevant and accurate information, one is able to choose in light of such information. In the case of breast cancer, there will be market forces pulling a breast cancer patient in one direction or another. Breast health centers, breast cancer specialists, and hospitals compete for breast cancer patients through media campaigns and attention to the services that they offer. In the case of reconstructive surgery, there will be market forces encouraging women to choose one reconstructive surgeon and surgery over another. In the case of the "pink kitsch," there is little we can do about the commercial products available online and in the hospital gift store, short of not buying them for oneself or others. But as long as the patient has been given the relevant information and can continue to choose within the context of the situation, her choices are free and there are no violations of her autonomy (for more on autonomy, see Chapter 7). Nevertheless, there are good reasons to be cautious about market forces in breast cancer care.

As we have seen, breast cancer is nested within frames of evaluations and social contexts replete with "nonmedical" influences. Another one of these nonmedical influences is politics, a topic that is considered next.

POLITICALIZATION

There is an initial tendency to view breast cancer as immune from political forces. We tend to think that our understanding of breast cancer is framed in terms of scientific and medical facts and not political claims. Of course, one would expect this in medicine. When one goes to the doctors, one expects a clinical response to what one brings to the attention of one's clinical professionals. One does not expect a homily, sales pitch, or political speech. One expects scientific thinking set within an asociological view of the clinical problem. In the case of breast cancer, one expects the current clinical evidence to be presented objectively and neutrally by a trained clinical professional licensed and certified in her or his field in breast cancer medicine.

In philosophy, the view that knowledge claims are apolitical is what I call the *apoliticalization of knowledge*. Such a view is rooted in the thinking of the early-twentieth-century logical positivists which endures today. Such thinking endures because it transcends particular communities of thinking and allows for cross-cultural analyses and discussions about natural phenomena. As discussed in Chapter 4, logical positivists hold that knowledge is arrived at by an analytic or empirical method that isolates, as Rudolf Carnap (1891–1970) (1966) says, "theoretical" or "observational" claims of knowledge. For logical positivists, knowledge is based on either statements of definition or empirical verifiability or falsifiability as established by scientific methods (Ayer 1952). All other ways to establish knowledge, including emotion, personal preference, and political opinion, are misleading and do not lead to knowledge.

In philosophy of medicine, the apoliticalization of knowledge is supported by what is called "evidence-based" medicine. Evidence-based medicine is "the use of mathematical estimates of the risk of benefit and harm, derived from high-quality research on population samples, to inform clinical decision-making in the diagnosis, investigation, or management of individual patients" (Greenhalgh 2010, 1). While its focus is on the best evidence in making decisions, evidence-based medicine can and does address more individual factors, such as quality-of-life and value-of-life judgments, which can be studied in controlled experimental settings (Sackett 2005). In such studies, evidence-based medicine relies on qualitative estimates of medical evidence or knowledge, which is seen to be separate and distinct from political influences.

Yet breast cancer is not apolitical. Political forces play a key role in fashioning how we understand and treat breast cancer. Knowledge of breast cancer is arrived at by an empirical method that isolates claims set within contexts of knowing, including political ones. One recalls the laws and policies mentioned in Chapter 1, which include the Women's Health and Cancer Act of 1998, the Breast and Cervical Cancer Prevention and Treatment Act of 2000, and the Affordable Care Act of 2010. Such laws and policies support and sustain certain views of and practices concerning breast cancer. Overt and "camouflaged" political forces frame our understanding and treatment of breast cancer. Medicine operates in social structures where politics are endemic. Politics guides who gets to decide and in what ways the allocation of resources in a society, including its laws and regulations. It guides what level of social support and funding will be required or recommended. One recalls from Chapter 1 the large amount of public funding that makes possible breast cancer care and research. Politics entails significant social control, which is typically justified by efforts to restore and preserve health and prevent against disease and illness (Foucault 1973 [1963]); Bartsky 1988; Morgan 1991, 86). In the case of breast cancer, widespread political support for breast cancer screening, testing, and intervention makes possible the kind of publicly funded breast cancer screening programs seen in the United States and elsewhere.

In philosophy, the view that our knowledge claims are influenced by political forces is what I call the *politicalization of knowledge*. By political, I mean of or relating to the government or public affairs of a state, nation, or country. In addressing the nature of scientific knowledge, feminist philosopher Elizabeth Anderson states that it is "impossible for individuals to rely only on themselves, for the very reason and interpretations of their experience on which they rely . . . is a social achievement" (1995, 53). In the context of science, and by extension medicine, "[i]ndividuals must use tools, methods, and conceptual frameworks developed by others in order to get their own theories under way. They must rely on the testimony of others to get evidence that is too costly or difficult for them to gather on their own, and even to interpret the evidence of their own senses" (Anderson 1995, 37). While Anderson does not address politics per se as a force that brings about knowledge, politics is included in that politics is one aspect of a human's so-called social achievement mentioned earlier. On this view, knowledge does not simply occur "between the ears"; it occurs within the human contexts and the historical, social, political, and cultural framework within which humans live (Jasanoff 2004).

In philosophy of medicine, the topic of the politicalization of cancer receives in-depth coverage from historian of science Robert Proctor. Proctor (1995) illustrates how government regulatory agencies, scientists, trade associations, and environmentalists have managed to obscure the issues and prevent efforts in medicine to enhance human welfare. He explores the association between the trade associations of the tobacco, meat, chlorine, and asbestos industries and scientific research projects to create and sustain uncertainty over the carcinogenic effects of their products. One consequence of this association is a restriction on the availability of knowledge about the carcinogenic effects of products on human health and the production of cancer.

In the case of breast cancer, Proctor offers a disturbing insight about medicine's approach to breast cancer. He says that "the hype surrounding the disease is used (by diagnostic firms promoting mammograms, for example) to further medicalize the female body, encouraging screening, but not prevention" (1995, 255). As an example, the British multinational pharmaceutical company, Zeneca (now called Astra-Zeneca), is both the manufacturer of the world's best-selling cancer drug, Tamoxifen (nolvadex), with sales of $470 million per year in 1995 (and sales of $1,024 million in 2001 ["Cancer" 2008]) and the founding sponsor of the National Breast Cancer Awareness Month in October in the United States with its annual pink media barrage urging women to have their annual mammograms (Ehrenreich 2001). Given that Astra-Zeneca makes significant profit selling Tamoxifen to women who have positive diagnoses for ER+ carcinoma in situ or invasive breast cancer (which accounts for 80% of breast cancer cases), and all US states support some version of a law regarding breast cancer screening and testing, it is difficult to argue that breast cancer is immune from political and economic forces (also see Batt 1994). One might also note the interest on the part of Astra-Zeneca and some breast specialists to encourage clinicians to prescribe Tamoxifen to women who have a family history of ER+ breast cancer (Grady 2013). One might wonder, with Lerner, if all of this emphasis on treating and preventing breast cancer with drugs distracts from the pursuit of knowledge about the carcinogenic effects of products on breast health: "as some activists have charged, is all of the attention of breast cancer screening and treatment distracting American society from pursuing a more productive goal—primary prevention of the disease by eliminating environmental and other toxins?" (2001, 13). The bottom line is that as much as we may think that breast cancer is immune from political forces, it is not.

CLOSING

Philosophically speaking, the question about how breast cancer is socialized raises a host of issues, including ones about the collaboration, medicalization, commodification, and politicalization of breast cancer. What we find initially is that our understanding of breast cancer appears to be insulated from social, nonmedical, economic, and political forces. But things are not as simple as one would think. Upon reflection, breast cancer is a collaborative endeavor, one framed by formal and informal social forces. In addition, breast cancer is medicalized, commodified, and politicalized. The medicalization of breast cancer is found in cases of labeling a breast phenomenon as "carcinoma" or "cancer" when it is not. The commodification of breast cancer comes about through endowing it with economic value and transforming aspects of it into an object that can be exchanged in the market. The politicization of breast cancer comes about through the events and activities of political agencies that rally on behalf of certain understandings and treatments of breast cancer.

Chapter 6

What Is the Relation Among the Descriptive, Explanatory, Evaluative, and Social Dimensions of Breast Cancer?

PERSONAL MUSINGS

As a breast cancer patient, I was often struck with how medicine separates what is not so separable. While reading my clinical and pathology reports, I attended carefully to the claims and data that were reported. I researched what was presented to me and tried to understand it and its implications for my care. In reading the reports and related literature, I often found that my attention moved away from the descriptions and explanations of my breast cancer to what this information meant to me, my loved ones, and to others in communities in which I participate. In other words, from my standpoint, the facts in the clinical and pathology reports were not just clinical facts; they were claims about pending harm, and they pointed in the direction of what ought to be done in order to return me to a state in which I would not face the risks that I did prior to being treated. Further, the descriptions and explanations in my clinical and pathology reports were not just about me; they were claims about medical standards in

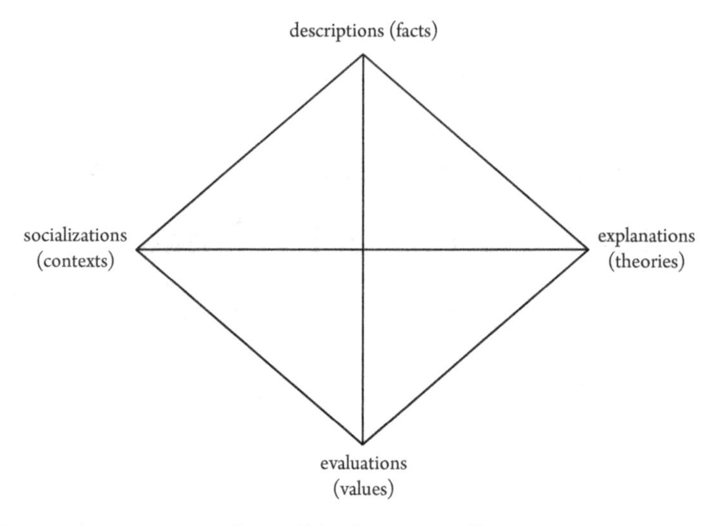

Figure 6.1 Interconnectedness of the dimensions of breast cancer.

breast cancer medicine and legal and policy standards adopted by my insurers and government. Still further, they were claims about my family, ones that carry significant implications for the females in my family line for an increased susceptibility to breast cancer. And so I asked, what is the relation among the various interests at play in understanding breast cancer? What is the relation among the facts, theories, values, and social interests that frame breast cancer? Herein lay the basis for some of my philosophical reflections on breast cancer as an integrative medical phenomenon (see Figure 6.1) and the need for a more personalized or precision-based approach to understanding and treating breast cancer.

AN INTERPLAY

As much as we may want to separate out the various dimensions of disease, the descriptive, explanatory, evaluative, and social dimensions are necessarily connected (Engelhardt 1996, 196). Consider the interconnectedness among the dimensions of breast cancer presented in Chapters 2–5. In this chapter, I draw from the earlier analyses and extend them to illustrate the intersections. To begin with, descriptions of breast cancer are composed of claims based on empirical evidence framed within prevailing contexts of reference through which such evidence is seen and interpreted. Currently, breast cancer is seen to be composed of mutated cells that result from abnormal processes that affect biological function in certain ways. Such mutated cells vary, and they may include ductal, lobular, or another kind of cell. Breast cancer may or may not be symptomatic, but if left alone, it will probably express signs and symptoms that lead to problematic outcomes for patients. Whether breast cancer is primarily physiological, genetic, immunological, and/or environmental is the focus of current research, which brings about new knowledge as new descriptions are forwarded.

Descriptions of breast cancer are nested in the explanatory theories through which they are seen and interpreted. Explanations of breast cancer bring coherence to the multiplicity of events that define breast cancer. They bring coherence to the signs and symptoms that bring patients into the clinic, the pathoanatomical and pathophyiological data that are generated by the laboratory, and the history of interpretation of a clinical condition. Currently, explanations of breast cancer rely heavily on etiological or causal accounts as they attempt to connect that which brings various kinds of breast cancer about. For instance, we know that the majority of cases of breast cancer arise from a defect of estrogen receptors in the breast tissue; fewer cases arise from isolated genetic mutations. As is the case for descriptions of breast cancer, explanations of breast cancer are nested within broader ontological and epistemological frameworks which themselves are open to change as clinical knowledge evolves.

Alternatively, explanations of breast cancer are nested in descriptive accounts through which they make sense out of abnormalities of the breast cells. They highlight the relations between and among the descriptive evidence that defines breast cancer. For instance, the discovery of ER+ breast cancer called for an explanation of breast cancer at the level of cell receptors. The discovery of BRCA1 called for an explanation of breast cancer at the level of the gene. Research on the environmental causes of breast cancer calls for an explanation of breast cancer attending to the patient's living environment, including air, water, and food. When the descriptions of breast cancer change, so do the explanations, as seen when new evidence emerges for distinct types of breast cancer. An example here is inflammatory breast cancer, which was once thought to be a result of contagion because it resembles an infection, but now is understood as the result of breast cancer cells blocking the lymph vessels in the skin (American Cancer Society 2015d).

Descriptions and explanations of breast cancer are set within frameworks of evaluations or norms. They are set within frameworks of evaluations of biological function, utility, beauty, and praiseworthiness. Such frameworks draw attention to what we as humans desire, prefer, like, and/or praise in the context of medicine. The values that frame disease function prescriptively; they serve as the basis for judgments about what is of worth. Norms of function and goals, and those of form and grace, are brought to bear on such judgments. For instance, norms of breast structure and function, and form and beauty, all enter into how breast cancer is understood. Recall the discussion in the previous chapter about the commodification of the female body. Further, clinicians and patients alike decide how to act, what to strive for, and what to resist in light of such norms. Breast cancer is a cluster of characteristics and abilities that serve as standards by which individuals and their conditions are judged to be "good" or "bad," or "right" or "wrong," instances of particular ideals of human function, goals, beauty, and choice. In this way, breast cancer is a normative phenomenon focused on how best to treat affected patients.

Alternatively, norms give rise to clinical descriptions and explanations. What and how humans value and disvalue influence what presents in the clinical curricula and textbooks, what is funded in clinical research, and what becomes foci of attention in local, regional, national, and international health policies. Individuals and communities rally around what is seen to be worthwhile targets of attention in health care, thereby encouraging the allocation of time, talent, and funds toward particular clinical endeavors that can result in new knowledge in clinical medicine. As an example, breast cancer care and research have received favorable attention during the last few decades and have resulted in the kind of funding that makes research on a wide scale possible. At the same time, when allocations are forthcoming for certain endeavors, individuals and communities do not pursue other avenues of pursuit, thereby discouraging time, talent, and funding toward other knowledge endeavors in clinical medicine. As an example, widespread funding to investigate the environmental contributions of breast cancer care has been slow in coming, thereby influencing how breast cancer is understood and treated. Either way, evaluations guide the descriptions and explanations health professionals adopt because they frame what actions are or are not worthy of attention, funding, and action.

As seen here, the descriptive and explanatory levels of analysis of breast cancer intersect with the prescriptive. Observations in medicine are ordered around theoretical commitments, including those concerning how to select and organize evidence into descriptions and explanations. Further, observations and explanations are ordered around evaluative commitments, including those concerning what phenomena are assigned significance in terms of what actions are seen to be appropriate in order to achieve certain goals. Observations and explanations in medicine are themselves instruments of human adaptation, and the value of cognitive claims is judged in terms of their utility, their adaptive value, and their ability to satisfy human needs and preferences. In the case of breast cancer, minimizing, if not eliminating, factors undermining patient welfare is central. Put another way, breast cancer serves as a treatment warrant that seeks to obtain particular clinical goals tied to enhancing a patient's welfare. Knowing clinical reality is never a purely theoretical endeavor. It involves practice or, as philosopher Marx Wartofsky (1976, 188) calls it, *praxis*, an endeavor of knowing and valuing, and doing and making. It involves "rich, historical contexts of fundamental and even revolutionary modes of cognitive praxis" (Wartofsky 1976, 188).

Given this, the descriptive, explanatory, and evaluative dimensions of breast cancer are nested in particular social contexts. A designation of breast cancer takes place within the social practices of developing professional standards, devising educational requirements and licensure agreements, formulating funding and business options, and instituting health laws and policies. To claim that a patient has breast cancer is to cast her in a "social role" (Parsons 1951) or as a member of the "kingdom of the sick" (Sontag 1978, 3) where certain societal expectations are forthcoming. Some of the societal expectations include assigning breast cancer patients a sick role, expecting that such patients seek help from specialists certified or licensed in their fields, excusing breast cancer patients from responsibilities for certain tasks when they undergo treatment, expecting that breast cancer care and research are a funding priority, expecting that breast cancer care is covered by medical insurance plans, and ensuring that breast cancer patients are protected against discrimination.

The social nesting of breast cancer has its challenges. There are aspects of breast cancer that are medicalized, commodified, and politicized. Breast cancer is medicalized when conditions that are not clinical conditions come to be seen as clinical problems. Diagnosing precancerous conditions such as ductal carcinoma in situ (DCIS) as "carcinoma" is an example. Breast cancer is commodified when conditions that are not usually exchanged become objects to be traded in market systems. Developing markets for breast screening, diagnosis, and treatment is an example. Breast cancer is politicized when government decides to regulate medical practice and research. Mandating insurance coverage of breast cancer screening, diagnosis, and treatment is an example. As seen here, the collaboration, medicalization, commodification, and politicalization of breast cancer can be problematic when patient welfare is compromised, and they can be beneficial when patient welfare is enhanced.

The multiple dimensions of breast cancer do not operate alone but rather interplay in various and complex ways. The descriptive, explanatory, evaluative, and social dimensions of disease define as well as situate each other. By "define," I mean form or mark the meaning of. By "situate," I mean locate, limit, or place within a certain place, position, or context. Descriptions define and situate explanations, as in the case of current causal explanations of breast cancer. More specifically, genetic descriptions give meaning to and locate genetic explanations of breast cancer within select interpretations and treatments of breast cancer. Alternatively, a lack of description of breast cancer situates or limits the kinds of causal explanations that can be offered in breast cancer medicine. For instance, our inability to describe the mechanisms of cellular dysfunction operating in particular kinds of breast cancer limits our ability to understand why certain types of breast cancer are more aggressive than others and why certain patients die from a specific kind of breast cancer while others do not.

Further, descriptions and explanations define and situate evaluations of breast cancer. Descriptions and explanations of breast cancer have led to successful treatment options for certain kinds of breast cancer. For instance, descriptions and explanations of HER2+ breast cancer made possible the development of the targeted treatment of Herceptin. Alternatively, a lack of descriptions and explanations for breast cancer limits treatment options because it limits what is studied and acted upon. For instance, a lack of knowledge about the distinct pathways of different kinds of breast cancer limits the development of targeted therapeutic options for particular kinds of breast cancer. Similarly, a lack of knowledge about pathways of different kinds of breast cancer limits the development of alternative treatment responses in breast cancer care.

Descriptions, explanations, and evaluations define and situate social expectations and actions. Descriptions, explanations, and evaluations of breast cancer have led to calls for more generous coverage of more accurate screens and tests for breast cancer patients. For instance, our understanding of BRCA1 and BRCA2 has set forth an expectation that genetic tests will be made more widely available in the United States and elsewhere to those who have a family history of breast cancer. Alternatively, the lack of descriptions, explanations, and evaluations of breast cancer limits what is supported on a social scale. For instance, there is need for greater attention to the customs, rituals, and behaviors that contribute to the rise of breast cancer rates in the twenty-first century. There is also need for continual reevaluation of the laws and policies that govern practices in breast cancer medicine in light of new descriptions, explanations, and evaluations of breast cancer.

Finally, social forces define and situate descriptions, explanations, and evaluations of breast cancer. Social forces contextualize the facts, theories, and evaluations of breast cancer. Social forces make possible large-scale research projects that generate new facts and theories about breast cancer within clinical communities governed by particular health policies and laws. For instance, consider the influences that advocacy groups such as Susan G. Komen (2014) and breastcancer.org (2015) have had on breast cancer research, screening, testing, and care. Alternatively, a lack of social involvement can deter the development of descriptions, explanations, and evaluations of breast cancer, thereby stalling the rise of new knowledge about breast cancer. For instance, a lack of attention to differences in breast cancer care across race/ethnic, class/economic, and age lines leads to an absence of knowledge about the "choice situations" (Sherwin 2006, 5) women face in breast cancer medicine. In short, the various dimensions of analysis of breast cancer are not so separate and distinct. The facts of breast cancer are theory laden, the fact–theory dyads are evaluative, and the fact–theory–value triads are socially framed. The various dimensions of breast cancer interplay in significant ways, a topic that is pursued in a more practical manner in the next section.

BREAST CANCER AS AN INTEGRATIVE MEDICAL PHENOMENON

Given the interplay among the descriptive, explanatory, evaluative, and social dimensions of breast cancer, breast cancer can be seen to be an example of an integrative medical phenomenon. Here I use the term "integrative" to show support for the important work being done by clinical practitioners in an area called "integrative medicine." Generally put, an integrative medical approach to disease challenges the view that a medical phenomenon is reductionistic, objective, determinist, and positivist (Koopsen and Young 2009, xvii). Here, for the integrative medical practitioner, "reductionism" occurs when complex phenomena are explained by component phenomena, as seen in the case of understanding breast cancer simply as uncontrolled cellular growth (see Chapter 2). "Objectivity" entails that the observer is separate from the observed and the observer can fully know the observed, as in the case of seeing a patient's disease as distinct from the patient, the disease and the patient as "objects," and the health care provider as an objective observer of the patient (see Chapters 2 and 3). "Determinism" occurs when phenomena are predicted from knowledge of scientific laws and initial conditions and given more predictive power than is warranted, as suggested in the phrase heard often in the media, "the gene for breast cancer" (see Chapter 3). "Positivism" means that information is derived strictly from physically measurable data, as is the trend in contemporary medicine with its reliance on laboratory data as opposed to the patient experience in understanding a disease finding (see Chapters 2–4.).

Put positively, an integrative approach to disease entails a balanced, whole-person-centered approach to health care and involves a synthesis of conventional or allopathic medicine, alternative or complementary modalities, and/or traditional medical systems, with the aim of prevention and health as a basic foundation. Here, an integrative approach to disease is to be distinguished from nonsynthetic approaches such as "alternative," "complementary," "unorthodox," "holistic," and "traditional folk" medicine that may seek to operate alone separate from conventional or allopathic medicine (Kligler and Lee 2004, 10; Koopsen and Young 2009). An integrative approach to disease recognizes that good medicine needs to be evidence based, be inquiry driven, and be open to new paradigms, but not without acknowledging the role patient reporting plays in guiding diagnosis and treatment. In this way, an integrative approach to disease rejects a reductionist, objective, determinist, and positivist approach to disease (Lee et al. 2004; Arizona Center 2009).

An integrative approach to breast cancer involves an integrative view of breast cancer. An integrative view of breast cancer comes about when the dimensions of description, explanation, evaluation, and socialization in breast cancer define and situate each other. As seen in the previous section, the descriptive, explanatory, evaluative, and social dimensions of breast cancer give meaning to and limit the clinical concept of breast cancer. While it may be difficult to pinpoint exactly how the interaction, articulation, and simultaneity of description, explanation, evaluation, and socialization frame breast cancer, we can begin to advance analyses and examples to think and act integratively. While traditional methods of understanding breast cancer focus on independent, discrete variables, an integrative model of breast cancer provides accounts that illustrate relationships between and among the processes and meanings of the dimensions. These include how they are created, maintained, changed, and critiqued (Mullings and Schultz 2006, 7), as is shown in this analysis of breast cancer.

Elsewhere, my colleague Raphael Sassower and I (2007) offer an integrative model of medical phenomenon in our discussion of the relation among diagnosis, treatment, and prognosis. As we say, "[d]iagnosis, treatment, and prognosis are at once both intimately related and independently derived" (Sassower and Cutter 2007, 86). As an example, "Whether a physician 'knows' the diagnosis, applies the 'appropriate' therapy or achieves the 'optimal' outcome, the patient as a 'closed system' will have manifested all of these elements as a unique unit" (Sassower and Cutter 2007, 86). For the patient, the diagnosis, prognosis, and treatment serve as a basis for explanation and evaluation within a social context. That which is diagnosed entails a future that serves as a basis for treatment and action—all set within the bodily and lived experience of a patient. For the patient, the descriptive, explanatory, evaluative, and social dimensions of disease are already intertwined. As Sassower and I say, "each component is dependent on the others and is intimately affected by them as well" (Sassower and Cutter 2007, 86).

In the case of breast cancer, when a breast cancer specialist "knows" she has detected Stage IIA invasive ductal carcinoma of the breast, she arrives at a certain prognosis and may recommend a lumpectomy or mastectomy with treatments, such as chemotherapy, radiation, adjuvant therapy, breast reconstruction, physical therapy, stress-reducing activities, nutritional counseling, and psychological counseling. She does this within a particular social context framed by expectations, responsibilities, and boundaries. Follow-up appointments and tests determine whether the prognosis and treatment are successful, which shed light on whether the diagnosis was accurate. With both a lumpectomy or mastectomy, pathological tests determine what was going on in the breast to confirm or change the diagnosis. In this process, the prognosis for the patient becomes more evident and treatment recommendations may be altered. While clinicians can and do separate diagnosis, treatment, and prognosis of a clinical problem, diagnosis, treatment, and prognosis are intimately related because the patient is a unique and closed biopsychosocial being with a particular expression of breast cancer and ways of responding to breast cancer treatment.

According to an integrative approach to a clinical phenomenon, the diagnosis, prognosis, and treatment of a clinical problem are fluid. "The treatment itself is . . . evaluated in comparison to the diagnosis so that the diagnosis itself can be revised, if needed, or maintained, if the results prove successful. Obviously, the diagnosis becomes a moving target in light of the results of the therapies or treatments and as such could inform the clinicians about the accuracy of the original diagnosis" (Sassower and Cutter 2007, 87). In the case of breast cancer, a breast cancer specialist may at first think that she has located a Stage IB ductal carcinoma, say a 1.8 cm tumor. Pathological tests may in turn show that the tumor is larger than initially detected, say a 2 cm tumor, and there may be evidence of metastases in the lymph nodes. If this occurs, the diagnosis will change from a Stage IB to a Stage IIA cancer, and treatments are revised in light of what is found. While this may appear at first to look like a clinician's ineptitude, it is not. It is the nature of clinical knowledge and the evolution of figuring out a clinical condition in light of situated and evolving knowledge that diagnostic and treatment recommendations change. Such is the kind of insight an integrative account of medical phenomena provides, peppered with a healthy sense of skepticism about clinical knowledge (see Chapters 3 and 4).

A practical way to understand an integrative account of disease is through the analogy of feedback loops. Sassower and I developed this account that might be of help:

> There is an epistemological and practical loop that encircles the diagnosis, treatment, and prognosis of disease. We become informed from one stage to another, recalling that which has been successful or failed to bring about anticipated outcomes. We collect data all along so that at each given moment we refer to our ongoing quest for greater knowledge accuracy and efficacy in treating patients. (Sassower and Cutter 2007, 95)

The information loop or cyclical feedback that Sassower and I (2007, 96) support has to do with breaking down the standard medical textbook view that one thing necessarily follows from another, that a patient contracts this or that disease by doing this or that, or that this particular gene is responsible for this particular disease. This should come as no surprise because medicine deals with physiological systems that transfer information among the components of all the systems. As philosopher Edward Yoxen puts it, "Nature is a system of systems. Organisms function, reproduce and evolve as systems ordered by their genes, 'managed' by the programme in their DNA. Life is the processing of information" (1983, 15). Genes operate in cells, tissues, and organs, which operate in intact bodies of persons who live their lives in particular environments. Life is a complex organic system of events and processes, the components of which work together in various and complex ways, not only in the patient's efforts toward health but in the conditions we call disease.

An integrative approach to breast cancer provides practical guidance in navigating the interrelations among the diagnosis, prognosis, and treatment of breast cancer. A decision has to be made by breast cancer specialists regarding how many cells with deviant changes of what kind in the breast tissue must be present before the cells are labeled as "cancer" (Cutter 1992). An explanation is given about the relation between the mutating cells and the result called "cancer." To be too liberal in classifying cells as "suspicious," "carcinoma," or "cancer" will lead to unnecessary anxiety, tests, and medical interventions, which harm women, cost money, and waste resources. To be too conservative in classifying cells as "suspicious," "carcinoma," or "cancer" will lead to a failure to diagnose a condition that is treatable, thus leading to premature death. How one discovers and creates lines among "suspicious," "carcinoma," and "cancer" involves appeals to facts, theories, values, and social considerations. How one discovers and creates lines among claims about "five-year" and "ten-year" survival, and among "progress" and "cure," also involve appeals to facts, theories, values, and social considerations. The facts of clinical and pathological investigation, theories of cancer causation, values of respecting patients and maximizing patient welfare, and social contexts of knowing and doing all play a role in guiding and sustaining particular diagnostic, prognostic, and therapeutic practices in breast cancer care. In this approach, conventional, alternative, complementary, unorthodox, holistic, and traditional folk approaches to understanding and treating breast cancer all find expression.

BREAST CANCER AS AN INTERSECTIONAL AND CONTEXTUAL PHENOMENON

An integrative account of disease is, philosophically speaking, an intersectional one. The way of thinking called "intersectionality" arises out of the work of feminists over the last few decades (Crenshaw 1989, 1991; Cornell 2004). It focuses on the relation among dimensions of influence and difference, such as gender/sexuality, race/ethnicity, and class/economics, in individual lives, social practices, institutional arrangements, and cultural forces. It attends to how these intersections lead to outcomes of intersections in terms of power. Kimberlé Williams Crenshaw (1989) introduced the idea of "intersectionality" to demonstrate how such dimensions frame the lives of women, and particularly black women, and the discrimination and oppression that they experience (Cornell 2004, 38). According to feminist sociologist Kathy Davis, intersectionality has two distinct characteristics. First, intersectionality understands "effects of race, class, and gender on women's identities, experience, and struggles for empowerment" (Davis 2008, 71). Second, it assists in deconstructing the "binary opposition and universalism inherent in the modernist paradigms of Western philosophy and science" (Davis 2008, 71).

Applied to our understanding of disease, dimensions of influence and difference (e.g., gender, sex, race, ethnicity, class, religion, and age) conveyed by descriptions, explanations, evaluations, and social framings of disease generate, shape, and maintain the health and disease of societies, communities, and individuals in ways that go undetected if not studied (Weber 2006, 25). An intersectional study of disease reveals, as this project has shown, the effects of multiple dimensions of influence and difference on patients' identities, experiences, and struggles for empowerment. An example here is how descriptions, explanations, and evaluations of disease turn on their social framings, even when this may not be evident in how we think about disease. An intersectional study of disease also reveals explicit, as well as implicit or assumed, binary oppositions and universal paradigms found in philosophy. Examples here are found throughout the previous chapters as oppositions have been contrasted to show the assumptions and claims we make about disease.

Applied to our study of breast cancer, dimensions of influence and difference conveyed by descriptions, explanations, evaluations, and socializations of breast cancer generate, shape, and maintain how breast cancer is understood and treated. Consider a recent study that sheds light on the social factors that influence whether breast cancer patients receive a lumpectomy or mastectomy. As journalist Elaine Schattner (2015) reports, "The *JAMA Surgery* report [Lautner et al. 2015] reveals a positive, albeit slow-to-pick-up trend: between 1998 and 2011, the proportion of U.S. women with early-stage breast cancers who had lumpectomy increased from 48.2% to 59.7%" (2015, ¶1). While this is to be celebrated, especially by those who welcome breast-sparing surgery for breast cancer patients, the study finds some striking social and economic differences among breast cancer patients and their treatments. As the study reports, "women with breast cancer who have private insurance, are more educated, earn more, receive care at academic medical centers, and live near a radiation treatment facility are more likely to have a lumpectomy. Women in the Northeast were most likely to get this smaller surgery, followed by those in the West" (Schattner 2015, ¶4). In this way, dimensions of influence and difference shape the "choice situations" (Sherwin 2005, 5) that breast cancer patients face. In contrast, those who received a mastectomy had the following risk factors: "being uninsured, having Medicaid or Medicare (including women under 65 years, with Medicare); having low income; not having completed high school; living in the South, receiving care at a community hospital; and living more than 17 miles from a treatment facility (for radiation)" (Schattner 2015, ¶3).

An integrative or intersectional account of disease is a *contextual* one. As physician and philosopher of medicine Lawrie Reznek says about a contextual account of disease: "[w]hether a condition is a disease depends on what sort of organism has the condition, and on the relation of that organism to the environment" (Reznek 1987,

169). Breast cancer is a disease in a culture that has evidence for a certain type of biological abnormality in the anatomical area called the breast of a patient. It is a disease in a culture that does not have the ability to control the overgrowth, unlimited survival, and subsequent disease process of mutated breast cells in a patient. Early ductal carcinoma in situ (DCIS) is a disease in a culture that has the diagnostic framework and clinical technology to test for it. Breast cancer is a genetic condition in a culture that has the means to diagnose genetic conditions. The bottom line is that "[w]e cannot decide whether a judgment about disease-status is true without considering the relation of the condition to the organism, and the relation of the organism to the environment" (Reznek 1987, 169).

A contextualist account of disease is not a *relativist* one. Conceptually speaking, relativism advances the view that "truth is relative." If truth is relative, then the claim itself is relative, thereby undermining its ability to establish anything. Practically speaking, relativism fails to take into account what common sense tells us: that we do know and act as if there is understanding in our world. We rely on medicine to describe and explain breast cancer, however partially, and it is often successful in doing so. Many strides have been made in diagnosing and treating breast cancer in the last fifty years. When we find mistakes in medical thinking, we have ways of correcting and revising them, so as to be more accurate and successful in our attempts in knowing and acting in the future. Despite the limitations, we have a consistent and effective vocabulary to describe, explain, evaluate, and socialize breast cancer for purposes of diagnosing and treating it.

The integrative phenomenon that we call breast cancer is not, then, relative. "Breast cancer" is a term medicine uses for observed signs found in the breast and associated symptoms of a patient. Granted, in the twentieth century, it is also a term that refers to the asymptomatically ill who may one day develop the signs and symptoms associated with breast cancer. Either way, breast cancer is an evolving notion that provides structure and significance to clinical reality and patient narratives. In talking about breast cancer, then, one must identify the particular community of clinicians and scientists to which one makes reference. Is one talking about an account of breast cancer offered by Hippocrates or Mary-Claire King? Is one referring to ductal or lobular breast cancer? Is one treating early- or late-stage breast cancer, Stage I or Stage III breast cancer, a 1.8 cm or a 5 cm tumor, or noninvasive or invasive breast cancer? Is one presented with ER+, HER2+, BRCA2, or some other type of breast cancer? Is one talking about one or five affected lymph nodes, or micrometastases or metastases? Is one referring to a 30- or 60-year-old patient, a patient with a family history of breast cancer, or a patient from New York City, Shanghai, or Accra? The point is that context matters for the diagnosis, prognosis, and treatment of breast cancer. Breast cancer is a contextual notion, and one that is not relative and doomed to a hopeless relativity ontologically, epistemologically, or evaluatively.

BREAST CANCER AS A PERSONALIZED MEDICAL PHENOMENON

Given that breast cancer is an integrative medical phenomenon, it makes sense that a personalized medical approach should guide the diagnosis, prognosis, and treatment of breast cancer. Personalized or precision-based medicine is a medical model developed in the late twentieth century that proposes that medical decisions, practices, and/or products are tailored to the individual patient. In the mid-nineteenth century, physician Heinrich Romberg (1795–1873) addressed the need for a more personalized approach in medicine: "And we do not regard the mere placing of disease under this or that rubric as the final aim of diagnosis. . . . The most important thing

[that] remains [is] the determination of the degree that the individual human is injured by his malady, and which cause has produced the momentary disorder" (Romberg, as translated in Engelhardt 1981 [1975], 36). One notes here Romberg's emphasis on the need in medicine to understand why an individual patient has a particular clinical condition.

Romberg was well before his time when he suggested a more individualized approach in medicine because medicine had yet to develop the means to achieve personalized medicine. In the late twentieth century, advances in genetics have fueled the development of personalized medicine. In April 1953, scientists and geneticists James Watson (1928–present) and Francis Crick (1916–2004) presented evidence for the DNA-helix, which carries genetic information from one generation to another (Watson and Crick 1953). Large collaborative projects, such as the Human Genome Project and the Hap Map, have provided the means to understand the role of genes in the expression of human disease. Here, the Human Genome Project is the "big science" project of the US government in the 1990s and early 2000s that mapped and sequenced the human genes (Biological Sciences Curriculum Study 1992). The Hap Map is the international project in the 2000s that mapped the haplotype or set of tightly linked genetic markers present on the chromosomes of the human genome. These projects have led to significant advances in the diagnosis and treatment of disease. As biologist Nicholas Wright Gillham says, "Deciphering of the human genome coupled with the HapMap and ever-decreasing cost of genomic sequencing has brought us to a new era where remedies can be targeted evermore precisely against diseases" (Gillham 2011, 235).

Coupled with advances in molecular pathology, the genetic revolution contributes to the development of personalized medicine in a number of ways. Drawing on a summary provided by physician and geneticist Francis S. Collins (1950–present), (1) "complete genome sequencing for many of us will include the discovery of various factors that will play a role in predicting susceptibility to illness" (Collins 2010, 181), (2) "[p]rediction of vaccine responses will also become possible" (Collins 2010, 181), (3) "sampling of one's personal microbiome [i.e., genome operating within an environmental context] will become an important part of the diagnostic workup for certain diseases" (Collins 2010, 181), and (4) drugs will be prescribed "based on genomic insight about your genes and the genes of pathogens" (Collins 2010, 181). On this view, personalized medicine is not simply about discovering genes. It is about understanding the role genes play within the context of a patient's body set within the context of the patient's environment and history.

A personalized medical approach makes sense in breast cancer medicine. In the early twenty-first century, much progress in breast cancer diagnostics and treatment has been made by testing for biomarkers and genes to determine specific kinds of breast cancer and specific ways to treat it. An example of the use of personalized medicine involves testing for the BRCA1 and BRCA2 genes, which are found in hereditary breast-ovarian cancer syndromes (National Cancer Institute 2015a). Genetic tests for BRCA1 and BRCA2 are able to diagnose presymptomatic cases of breast cancer and allow individuals to grapple with tough decisions involving prophylactic measures, such as a mastectomy and removal of ovaries, to reduce the chance of expressing breast cancer. Recall the journey shared by actress Angelina Jolie (2013, 2015) when she underwent a prophylactic bilateral mastectomy after testing positive for BRCA1 and losing family members to related conditions. More detailed molecular tests of breast tumors beyond BRCA1 and BRCA2 will pave the way for future tailored treatments.

But keep in mind that personalized medicine is not simply about discovering genes. It is about understanding biological uniqueness and variability. As Vogelstein says, "[e]very patient's cancer is unique because every genome is unique. Physiological heterogeneity is genetic heterogeneity" (Vogelstein, as quoted in Mukherjee 2010, 452). Given this, more targeted therapies beyond lumpectomies and mastectomies, and the "one-size-fits-all" therapies commonly used on breast cancer medicine today are in order. An emphasis on biological uniqueness and variability calls "into question the logic of performing the same operation on all women with the disease" (Lerner 2001, 12). An emphasis on the biological uniqueness and variability of breast cancer also calls into question a heavy reliance on the TNM staging system of classification for breast cancer. Although this system is more available to clinicians worldwide, there are good reasons to move toward biologically based diagnoses of specific types of breast cancer in order to offer more specific types of treatment.

A personalized medical approach to breast cancer care makes philosophical sense because it emphasizes the integration and intersectionalism among the multiple dimensions of breast cancer. It emphasizes the relation of the biological expression of breast cancer to the individual patient, and the relation of the patient to his or her environment. Multidisciplinary thinking and multidisciplinary practice offer hope for knowledge about unique and different types of breast cancer, their prognosis, and their treatments. In this way, a personalized medical approach to breast cancer is an integrative, intersectional, and contextual one.

CLOSING

Philosophically speaking, breast cancer is an integrative phenomenon. The descriptive, explanatory, evaluative, and social dimensions of breast cancer mutually define and situate each other within particular contexts. Descriptions, observations, or the "facts" of breast cancer are ordered around explanatory assumptions and claims.

Further, the descriptive and explanatory dimensions of breast cancer are ordered around evaluative or prescriptive judgments, including those objects that are assigned significance and those actions that are seen to be appropriate and warranted in order to achieve certain goals. These goals include minimizing patient pain and suffering, returning a patient to biological functioning and well-being, honoring patient choices, satisfying professional interests and standards, ensuring financial stability, maintaining cultural and societal norms, and so on. Still further, the descriptive, explanatory, and evaluative dimensions of breast cancer are socially located. There is no timeless or noncontextual account of breast cancer, or at least there is no such interpretation available to humans. Breast cancer is an intersectional and contextual phenomenon. In this way, breast cancer warrants a personalized medical approach to its diagnosis, prognosis, and treatment.

Chapter 7

What Are the Ethical Implications of Understanding and Treating Breast Cancer?

PERSONAL MUSINGS

As a breast cancer patient, I often pondered the ethical implications of how medicine understands and treats breast cancer. I recognized at the start that the diagnosis of my breast cancer carried practical implications. I knew after my positive diagnosis of ER+ breast cancer that my life would not be the same. With such practical implications came a series of choices. There were choices on my part regarding consenting to a series of medical interventions. There were choices on my breast cancer specialists' parts to interpret findings and present the evidence to me in a certain way. There were choices on the part of those paying for my care to fund certain procedures and not fund others. Given that choice is a necessary condition of an ethical life, it struck me that the very diagnosis of my breast cancer carried ethical implications. Here "ethical" (from the Ancient Greek root *ēthikós,* meaning "of or concerning or expressing character") refers to having the quality of judgment concerning the praiseworthiness or blameworthiness of an action or character. Some of the ethical implications

autonomy: duty to respect self-determination
nonmaleficence: duty to minimize or avoid harm
beneficence: duty to benefit another
justice: duty to treat others fairly

Figure 7.1 Principles of biomedical ethics.

that I pondered during my time as a breast cancer patient focused on decisions to grant permission for my care, decisions to rank the benefits and harms of my medical interventions, and decisions regarding structuring access to breast cancer care in a certain way. In particular, I asked, how could I give informed consent for my breast cancer care when I clearly was not the clinical expert? What moral account of informed consent makes sense in a context in which breast cancer diagnosis and treatment are not certain? Given the uncertainty of breast cancer diagnosis and treatment, how does one begin to manage the risks of over- and underdiagnosing as well as over- and undertreating breast cancer? What are the ethical implications of decisions about risk assessment in breast cancer care? What constitutes proper or just access to breast cancer care? Herein lay the basis of some of my philosophical reflections on the ethical implications of how medicine understands and treats breast cancer and the ethical issues concerning informed consent, risk assessment, and just access to breast cancer care. Here, the principles of autonomy, nonmaleficence, beneficence, and justice in biomedical ethics are discussed (see Figure 7.1).

In addition, the ethical principles of autonomy, nonmaleficence, beneficence, and justice are critiqued in light of an integrative approach to how we understand breast cancer. The message is that how we understand and treat breast cancer carries ethical implications.

INFORMED CONSENT AND AUTONOMY

A breast cancer patient is faced with signing numerous consent forms that give permission for health care professionals to intervene in the clinical setting. Here "consent," from the Latin roots *con,* meaning "together," and *sentire,* meaning "thinking or judgment," means the possession or display of reliable information or knowledge. Informed consent is a legal procedure to ensure that a patient, client, or research participant is aware of the potential risks and costs involved in a medical procedure. Since the mid-twentieth century, medicine has incorporated the requirement to secure informed consent into its practice (Faden and Beauchamp 1986; Bowman et al. 2012). The concept and practice of informed consent develop hand in hand with the rise of the recognition of patient rights.

The patient rights movement has supported the view that patients have a right to be treated in a respectful way in the health care setting. This view is based on a *principle of autonomy* and instructs each person to acknowledge another's "right to hold views, to make choices, and to take actions based on their values and beliefs" (Beauchamp and Childress 2013, 106). The term "autonomy" comes from the Greek roots *auto,* meaning "self," and *nomos,* meaning "custom" or "law." The principle of autonomy is a moral and legal concept rooted in the thinking of philosopher Immanuel Kant (1985 [1785]) and his position that respect for person is a necessary condition of the possibility of the moral community. As Kant says in *Foundations of the Metaphysics of Morals*: "Act so that you treat humanity, whether in your own person or in that of another always as an end and never as a means only" (Kant 1985 [1785], 47). The principle of autonomy requires the treatment of another as an end in herself or himself and not as a means to some other goal. Here, an "end" refers to being treated as a being worthy of respect as an independent decision maker. Autonomy involves two conditions: (a) liberty, or "independence of controlling influences" (Beauchamp and Childess 2013, 102), and (b) agency, or "capacity for intentional action" (Beauchamp and Childress 2013, 102) (see Figure 7.2).

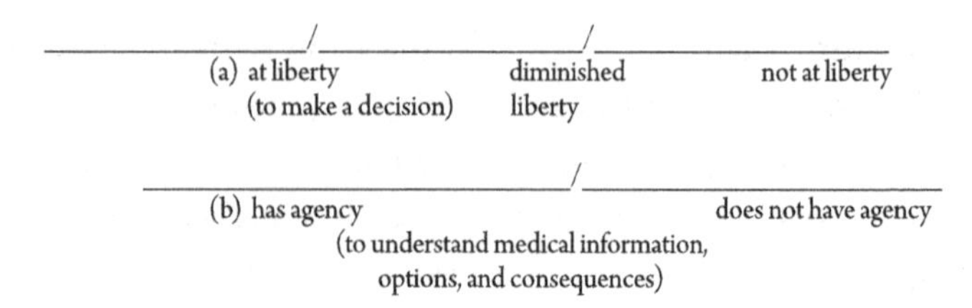

Figure 7.2 Spectrum of autonomy.

Autonomy requires the noninterference of another's attitudes and actions as well as an obligation to build up or maintain another's "capacities for autonomous choice while helping to allay fears and other conditions that destroy or disrupt autonomous action" (Beauchamp and Childress 2013, 107). Put practically, in medicine, autonomy grounds the practice of informed consent as well as truth-telling, respect for privacy, and protection of confidential information (Beauchamp and Childress 2013, 107).

The recognition of autonomy has benefited and continues to benefit breast cancer patients and research subjects. It requires acknowledging a breast cancer patient as an end in herself (and not the means to something else) as well as protecting a patient's capacity for autonomous decision making. It requires the avoidance of deception in breast cancer care and the provision of relevant information so that a breast cancer patient can and may make informed decisions about her care. In these capacities, it secures a zone of privacy for a breast cancer patient protecting her in her most personal health-related decisions. Such grounds the moral importance of informed consent in breast cancer care. Let's take a closer look at informed consent, the moral obligation to secure it, and the ethical challenges that emerge from such obligation in breast cancer care. What follows does not attempt to cover all aspects of informed consent. It rather provides a geography of issues, thus pointing in the direction of topics worthy of pursuit in breast cancer medicine.

I. Threshold elements (preconditions)
1. competence (to understand and decide)
2. voluntariness (in deciding)
II. Information elements
3. disclosure (of material information)
4. recommendation (of a plan)
5. understanding (of 3 and 4)
III. Consent elements
6. decision (in favor of a plan)
7. authorization (of the chosen plan) (Beauchamp and Childress 2013, 124)

Figure 7.3 Elements of informed consent.

Informed consent is a complex notion. According to bioethicists Tom L. Beauchamp and James F. Childress (1940–present), there are seven components of informed consent (see Figure 7.3).

As shown in Figure 7.3, informed consent in the health care setting entails (1) the competence to understand and decide, (2) a state of voluntariness in deciding, (3) the disclosure of relevant medical information, (4) the professional's recommendation of a relevant plan for treatment, (5) the ability to understand incoming information and to have relevant beliefs about the nature and consequence of the proposed treatment plan, (6) the ability to decide in favor or against a treatment plan, and (7) the ability to authorize a chosen treatment plan.

Applied to our discussion, informed consent in breast cancer care entails (1) the breast cancer patient's ability to understand her condition and decide her course of action, (2) a state of voluntariness or freedom in deciding her course of action, (3) the provision of relevant or material breast cancer information, (4) a breast cancer specialist's recommendation of a relevant plan of treatment, (5) the patient's ability to understand the incoming information and to have relevant beliefs about the nature and consequences of the proposed treatment plan, (6) the patient's ability to decide in favor of or against a treatment plan, and (7) the patient's ability to authorize a course of action for breast cancer care and proceed accordingly.

There are a number of conditions that enhance informed consent in breast cancer care. The threshold elements of a patient's (1) competence and (2) voluntariness are enhanced when a patient is sufficiently free from controlling influences in order to receive medical information, process it, and respond to it. The provision of (3) relevant medical information, (4) recommendation for treatment, and (5) a patient's understanding of the information and treatment plan are enhanced when the information and treatment plan are made available to patients in a way that is conducive to receiving and processing it. Given the large number of specialists involved in breast cancer care, it is beneficial to have as many specialists meet with the patient altogether to ensure that the patient understands her clinical situation and the recommended treatment plan that is offered. Otherwise, when a patient meets individually with specialists, as is often the case in nonacademic medical settings, a patient finds herself having to report on what another specialist says and thinking about how each of the specialist's recommendations fits together into a coherent treatment plan. When a team-advising approach is used, the patient is better equipped (6) to make decisions and (7) to authorize a plan of action in light of her understanding and ability to act on it.

But even if these conditions are met, there are additional challenges with securing informed consent in breast cancer care. As we learned in the previous chapters, (1) the breast cancer patient's ability to understand her condition and decide her course of action, (2) the patient's freedom in deciding her course of action, (3) the disclosure of relevant medical information, (4) and clinical recommendations about a proposed plan of action are fraught with uncertainties. They are fraught with uncertainties because the medical information and clinical recommendations patients seek to understand are probabilistic; they are incomplete, temporary, and open to change (see Chapter 3). As social scientist Gerd Gigerenzer (1947–present) puts the general problem, "not everything [in medicine] is known as precisely as we would wish. The lifeblood of the illusion of certainty . . . is thinking in black-and-white terms rather than in shades of gray" (Gigerenzer 2002, 99–100). In light of this, consider Gigerenzer's advice about how to think about the problem. Choices about medical information, including breast cancer information, are not ones "between *certainty* and *risk*" but rather "a choice between risks" (Gigerenzer 2002, 99). In risk assessment, "[e]ach alternative carries its own uncertain consequences, which need to be compared for an informed decision to be made" (Gigerenzer 2002, 99). Given this, it follows that a breast cancer patient's understanding (5) will always be less than certain, and she will have to come to terms with the fact that she is (6) deciding and (7) authorizing a plan of action on incomplete, evolving information.

Given the limits of the certainty in breast cancer medicine, informed consent might better be labeled as "uninformed consent" (Gigerenzer 2002, 98). This is a tough conclusion for me to draw as a breast cancer patient and someone who has devoted her academic career to supporting the practice of informed consent in the clinical setting. But bear with me for a moment as I draw out Gigerenzer's points as they would apply in breast cancer care. Gigerenzer argues that "uninformed consent" comes about through a number of forces in medicine: division of labor, legal and financial incentive structure, conflicts of interest, and innumeracy (2002, 113–114). First, according to Gigerenzer, medicine is built on a division of labor and tests are performed by different professionals who do not always speak directly with one another. As an example, a radiologist who performs mammograms usually does not know which patients later develop cancer and which do not. Information about individual patient outcome may aid the radiologist in reading mammogram results, which are challenging to read anyway ("Radiologists" 2007). Second, there are significant legal incentives in medicine to perform tests to prevent against lawsuits. Performing more tests leads to "a large number of false positives and their physical, psychological, and monetary costs to the patient" (Gigerenzer 2002, 113). It is estimated that more than 50% of women in the United States who get annual mammograms will have at least one false-positive reading after ten years of screening (Boyles 2012). Third, policies and recommendations regarding medical tests are not just policies and recommendations; they carry financial implications. Encouraging women to be screened for breast cancer annually carries with it large financial advantages for breast cancer specialists and the companies that produce the diagnostic testing and pharmaceutical drugs to treat breast cancer (recall Proctor 1995, in Chapter 5). Fourth, "[m]any physicians are poorly educated in statistical thinking and have little incentive to engage in this alien form of reasoning" (Gigerenzer 2002, 114). And yet, medical information is built on risk assessments drawn from research studies. And then there is something that Gigerenzer does not mention: the constraints of time. How often is one asked at the beginning of a clinical appointment to sign a page with small print and lengthy text? Who has time to digest such information? Who has time to ask questions about what is contained in it during the limited time one has with a clinician? All in all, such forces undermine informed consent in medicine in general, and in breast cancer care in particular.

And then there is this: a growing body of literature on informed consent shows that the view of an autonomous agent in medicine is misleading. Decision makers in medicine work within communities of knowers and doers and decide directions of actions based on numerous and various forces, some of which are explicit and others more implicit. Breast cancer patients typically decide on courses of action within their communities of families, friends, and support groups. Breast cancer specialists work within communities as well. Both patients and specialists make decisions within a much wider context than represented by isolated autonomous decision makers. In this way, the dominant sense of autonomy that has been adopted in medicine today contradicts how patients actually make decisions. Understanding how breast cancer patients and their clinicians make decisions would make the informed consent process in breast cancer medicine a more fruitful, meaningful, and authentic process.

On this point, feminist philosopher and bioethicist Susan Sherwin addresses the need to reconceive autonomy, and by implication informed consent, as relational. This is in keeping with the call for an integrative or intersectional approach in breast cancer medicine (see Chapter 6). Sherwin argues that there are a number of problems with the current view of autonomy in medicine. In particular, "we need to question how much control individual patients really have over the determination of their treatment" (Sherwin 1998, 24). Health care specialists are not only medical authorities but bound by numerous professional and legal standards in their practice. Further, according to Sherwin, there are "deeper" problems. First, standards of autonomy lack "satisfactory guidelines" "when dealing with patients who do not fit the paradigm" (Sherwin 1998, 24). Think about it: none of us are rational, self-determining, free agents in the way often featured in accounts of autonomy. Second, "autonomy is often understood to exist in conflict with the demands of justice because the requirements of the latter may have to be imposed on unwilling citizens" (Sherwin 1998, 25). According to Sherwin, the conflict between autonomy and justice sets up a misleading contradiction because autonomous agents rely on others for information and access to services in order to be autonomous. Given this, autonomy and justice work together rather than in opposition to each other. Third, "autonomy is often used to hide the workings of privilege and to mask the barriers of oppression" (Sherwin 1998, 25). Here privilege is not simply a matter of economics, but of education, institutional power, and gender/sexual, ethnic/racial, class/economic, religious/spiritual, and age-related dimensions of difference. Rallying around autonomy can mask the institutional forces that lead patients to decide in the ways that they do and to be treated in the ways in which they are treated. Fourth, "the standard conception of autonomy . . . tends to place the focus of concern quite narrowly on particular decisions of individuals" (Sherwin 1998, 26), when in fact decisions in medicine are communal and involve others' input, influence, pressure, and support. The paradigm of a patient as a solitary, rational, free agent who decides in light of neutral and clear evidence does not fit clinical reality. In the end, there is need to conceive autonomy and informed consent as relational and contextual.

How does one begin to see autonomy and informed consent as relational and contextual? According to Sherwin, relational autonomy is "a capacity or skill that is developed (and constrained) by social circumstances. It is exercised within relationships and social structures that jointly help to shape the individual" (Sherwin 1998, 36). In order to achieve relational autonomy, we might begin to attend to the "power imbalances, especially those associated with oppression, [that] can skew the array of choices available to support the interests of the powerful while constraining the capacity of the oppressed to find options that reduce their oppression" (Sherwin 2006, 15). A close look at "whether adequate social conditions are in place to facilitate choices that support the interests of both the individual and whatever social groups she belongs to" (Sherwin 2006, 15) is a step that can be taken. Given that most medical options are "built around the metaphors of warfare and risk factors" (Sherwin 2006, 15) within a framework in which medico-scientific professionals have concentrated power, a look at alternative frames of understanding breast cancer and a reconsideration of the power relations between medico-scientific professionals and patients/consumers might be helpful. In addition, "more diversity in research approaches to breast cancer" (Sherwin 2006, 15) might "alter the pattern of incidence" (Sherwin 2006, 15) of breast cancer. Sherwin calls us to do some serious rethinking of how we understand and treat breast cancer: "If research into breast cancer were pursued under this model, women's knowledge would not be limited to what commercially driven science has seen fit to study. It would also reflect the range of questions that feminists raise about breast cancer, including questions about how to promote breast health, the need to limit industrial activities that pose a risk to women's health (Rosser 2001), the reasons for racial and ethnic differences in diagnosis and mortality (Kasper and Ferguson 2000), and the availability of strategies to live well despite the presence of breast cancer" (Sherwin 2006, 15). The challenge is daunting: "So powerful is the coalition of voices working within the dominant biomedical model that it is difficult for anyone—patient, physician, or researcher—even to imagine alternative understandings of the disease" (Sherwin 2006, 16). But imagine, we must. And if this fails to reap benefits, then we can reimagine another way.

All of this returns us to a theme explored in Chapter 6: the integrative character of breast cancer. With a better appreciation of how breast cancer patients and specialists make decisions in medicine, the achievement of the moral standard of respecting patient autonomy and securing informed consent may better be reached. Perhaps some lessons here may be drawn from how breast cancer patients and specialists make the decisions they do in light of risk assessments in breast cancer medicine, a topic that is considered in the next section.

RISK ASSESSMENT, NONMALEFICENCE, AND BENEFICENCE

A breast cancer patient is faced with numerous decisions regarding diagnostic procedures and treatments during her time as a patient. Such decisions typically involve considerations about determining and weighing benefits and harms, or what is called "risk assessment."

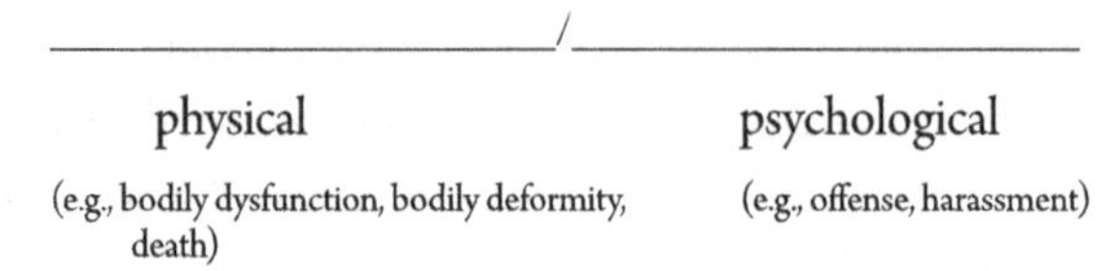

Figure 7.4 Spectrum of harm.

Here risk assessment "involves the analysis and evaluation of probabilities of negative outcome, especially harms" (Beauchamp and Childress 2013, 232). "Risk" "refers to a possible future harm" (230) and "harm" as a "setback to interests particularly in life, health, and welfare" (Beauchamp and Childress 2013, 230). Harm is a complex notion that involves considerations ranging from physical harm to psychological harm (see Figure 7.4).

Statements such as "*minimal risk, reasonable risk,* and *high risk* usually refer to the chance of a harm's occurrence—its probability—but often also to the severity of the harm if it occurs—its magnitude" (Beauchamp and Childress 2013, 230). In medicine, there is a longstanding moral obligation to minimize patient harm. Harm varies, as Figure 7.4 indicates. There are some harms that involve low probability and low magnitude. Others involve high probability and high magnitude, and others are some combination thereof. Harms that involve high probability and high magnitude are typically of great concern in medicine. Harms that involve low probability and low magnitude are typically of less concern in medicine. Yet there are great variations in between. Because of the variations in the probability and magnitude of the harms brought about by disease and its treatments, medicine has devoted much attention to the gathering of risk assessment data for particular medical interventions of particular diseases. Let's take a closer look at risk assessment in breast cancer care through the lens of the moral obligations to minimize patient harm and maximize patient welfare and the ethical challenges that emerge from such obligations. What follows does not attempt to cover all aspects of risk assessment. It rather provides a geography of issues, thus pointing in the direction of topics worthy of pursuit in breast cancer medicine.

On the one hand, there is the long-standing ethical mandate in medicine to "do no harm." In today's bioethical literature, this mandate is referred to as the *principle of nonmaleficence*. The term "nonmaleficence" comes from the Latin roots *non*, meaning "no," and *malificencia*, meaning "evildoing." The principle of nonmaleficence "obligates us to abstain from causing harm to others" (Beauchamp and Childress 2013, 150). It involves intentionally *refraining* from actions that cause harm, where "harm" is understood as a "thwarting, defeating, or setting back of some party's interests" (Beauchamp and Childress 2013, 153). The principle of nonmaleficence has come to be closely associated with the familiar maxim, *primum non nocere*, which means "above all [or first] do not harm," a notion rooted in Hippocratic thinking (Hippocrates 1923, 165). It supports specific moral actions in medicine such as refraining from killing, causing pain or suffering, incapacitating another, causing offense, and depriving another of the goods of life (Beauchamp and Childress 2013, 154).

In the case of breast cancer, the principle of nonmaleficence obligates a breast cancer specialist to abstain from causing harm to a patient. This means that the breast care specialist is to abstain from thwarting, defeating, or setting back the patient's interests. But anyone who has been through breast cancer treatment knows that many of the clinical interventions (e.g., lumpectomy, mastectomy, breast reconstruction, chemotherapy, radiation) typically involve harmful side effects (e.g., pain, scarring, neuropathy, loss of sensation, lymphedema, physical deformity, reactions to drugs). So, yes, breast cancer specialists can and do harm patients. The question becomes, then, how does one address the morality of such harms? Drawing from Gigerenzer (2002, 99), we are reminded that the choice is not between certainty and risk; it is a choice between risks. Further, it is a choice regarding the degree and magnitude of the risks. Given that for early-stage breast cancer, a lumpectomy is shown to be as effective as a mastectomy and a lumpectomy is less invasive than a mastectomy (Gottlieb 2000, 261), the principle of nonmaleficence obligates the breast cancer specialist to recommend the approach that is least harmful, and in this case a lumpectomy.

The principle of nonmaleficence operates as well in decisions about screening and testing for breast cancer. In the last decade, a debate has developed about the appropriate use of mammography screenings in breast cancer care (Plutynski 2012). The debate entertains a number of questions, including at what age and how often women should be screened for breast cancer. Nonmaleficence requires that health care professionals take the least harmful approach in offering such screenings and diagnostic to minimize the harmful effects of a mammography (e.g., exposure to x-rays) as much as possible. In following this mandate, the US Preventive Service Task Force recommends that women 50 through 74 years of age should be screened every two years (US Preventive Services 2009) and not every year as the American Cancer Society ("Breast Cancer Screening" 2015) advocates. The question becomes, "Should women under age 50 undergo routine screening mammography? Does finding a small number of cancers in such women justify the negative ramifications, such as unnecessary biopsies for those who are tested?" (Lerner 2001, 13).

Complementing the principle of nonmaleficence is the *principle of beneficence*. The term "beneficence" comes from the Latin roots *bene*, meaning "well," and *facio*, meaning "to do." The principle of beneficence "demands more than the principle of nonmaleficence, because agents must take positive steps to help others, not merely refrain from harmful acts" (Beauchamp and Childress 2013, 202). The principle of beneficence is in the tradition of John Stuart Mill's (1979 [1861]) ethical mandate to promote the best interest of another. As Mill says in *Utilitarianism*: "The creed which accepts as

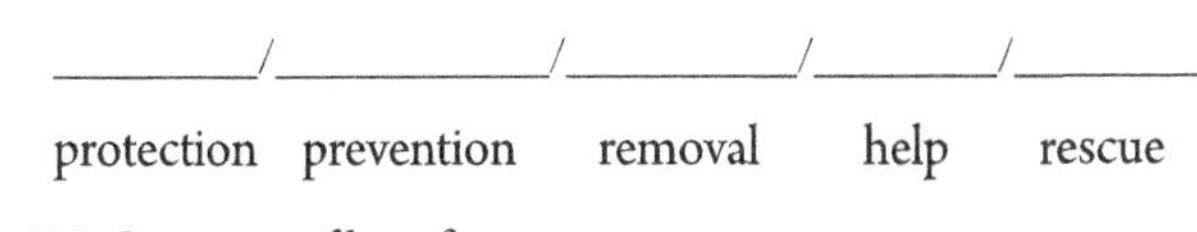

Figure 7.5 Spectrum of benefit.

the foundation of morals 'utility' or 'the greatest happiness principle' holds that actions are right in proportion as they tend to promote happiness; wrong as they tend to produce the reverse of happiness" (Mill 1979 [1861], 7).

The principle of beneficence supports *general* rules of moral actions that benefit others such as protecting and defending the rights of others, preventing harm from occurring to others, removing conditions that will cause harm to others, helping persons with disabilities, and rescuing persons in danger (Beauchamp and Childress 2013, 204) (see Figure 7.5).

Analogous to nonmaleficence, statements such as minimal risk, reasonable risk, and high risk refer to the chance of a benefit's occurrence—its probability—but also to the magnitude of the benefit if it occurs. The principle of beneficence obligates the clinician to maximize patient benefits in light of the probability and magnitude of their occurrence. Benefits vary, as Figure 7.5 indicates. There are some benefits that involve low probability and low magnitude. Others involve high probability and high magnitude, and others are some combination thereof. Benefits that involve high probability and high magnitude are typically more highly valued and sought in medicine. Those that are determined to be low probability and low magnitude are typically less pursued, unless this is the only option.

The principle of beneficence also supports *specific* rules directed at persons such as patients. Beauchamp and Childress argue that a person X has a specific obligation of beneficence toward person Y if and only if each of the following conditions is satisfied (see Figure 7.6).

1. Y is at risk of significant loss or damage to life, health, or some other basic interest.
2. X's action is necessary (singly or in concert with others) to prevent this loss or damage.
3. X's action (singly or in concert with each other) will probably prevent this loss or damage.
4. X's action would not present significant risks, costs, or burdens to Y.
5. The benefit that Y can be expected to gain outweighs any harms, costs, or burdens that X is likely to incur (Beauchamp and Childress 2013, 207).

Figure 7.6 Specific rules of beneficence.

Applied to the case of breast cancer, a breast cancer patient is at significant loss or damage to health, and perhaps life, if she does not receive medical care. Clinical action is necessary to prevent this loss or damage. A clinician trained in breast cancer intervention can in all likelihood prevent this loss or damage. Such intervention would probably not present significant risks, costs, or burdens to the patient, or at least the benefit that the patient can be expected to gain outweighs the harms, costs, or burdens that the clinician is likely to incur. Again, in the case of breast cancer, while a mastectomy is a harsh clinical intervention that can bring about short-term as well as long-term pain and suffering, the benefits of the intervention are calculated to outweigh the harms for certain breast cancer patients at certain stages of cancer. The same can be said about chemotherapy and radiation.

In thinking through how to determine what constitutes one's duty of nonmaleficence and beneficence, much turns on what constitutes harm and benefit to patients. In medicine, risk assessment attempts to clarify this. Here, informal and formal techniques for such risk calculations have been developed (Beauchamp and Childress 2013, 231). Informal techniques include "expert judgments based on reliable data and analogical reasoning based on precedents" (Beauchamp and Childress 2013, 230). These techniques are typically used in Institutional Review Board (IRB) protocols, where the investigator must state the risks to subjects and probable benefits to both subjects and society and justify how the probable benefits outweigh the harms to the subject. Formal, quantitative techniques involve the analysis of "the ratio between the probability and magnitude of an anticipated benefit and the probability and magnitude of an anticipated harm" (Beauchamp and Childress 2013, 230). Formal techniques often draw a distinction between absolute versus relative risk, where absolute risk in the context of breast cancer screening "is calculated by subtracting the mortality rate in the screened from the unscreened group" (Plutynski 2012, 11) and "[r]elative risk is calculated by dividing the decrease in morbidity or mortality by the baseline rate" (Plutynski 2012, 11). In other words, absolute risk indicates the probability that a person will develop cancer over the course of a lifetime. Relative risk compares the risk of developing cancer between persons with a certain exposure or characteristic and persons who do not have this exposure or characteristic. Determining risk for disease is a sophisticated body of techniques (and is too much to review here), but for purposes of illustration it involves cost-effectiveness analysis (CEA, which measures the benefits in nonmonetary terms such as years of life, quality-adjusted-life-years [QALYS], and cases of disease), cost–benefit analysis (CBA, which measures both the benefits and costs in monetary terms), and risk–benefit analysis (RBA, which evaluates risks in relation to probable benefits).

The moral imperative to minimize harms and maximize benefits in medicine is important in breast cancer care. This imperative has been grounds for more accurate risk assessments in breast cancer medicine. More accurate risk assessments serve as the basis for more tailored and personalized diagnostic tests and treatment plans for breast cancer (in the case of HER2+ breast cancer, for instance), which has benefited women by reducing patient pain and suffering, thereby helping patients to return to a desired level of function, preventing the onset of more advanced breast cancer, and reducing mortality from breast cancer. The moral imperative to minimize harms and maximize benefits in medicine has been grounds for campaigns for more research on breast cancer in order to reduce further the morbidity and mortality from breast cancer, a reevaluation of the use of certain pharmaceutical agents that carry significant side effects, and a move to replace mastectomies with lumpectomies and other clinical interventions in order to offer less invasive medical interventions that carry fewer risks. Such activities reflect a moral commitment to minimize harms and maximize benefits for breast cancer patients.

All sounds fine and good, but there are challenges with calculating risk assessments and applying this knowledge in breast cancer care. Aronowitz states the overall challenge this way: "Risk . . . is a cultural construct that bears a problematic and often indirect relationship to death rates or other 'objective' markers of danger and bad outcomes. In our contemporary response to breast cancer, risk is an elusive term with different meanings and uses" (2007, 5). Elaborating on the ambiguity, he says that risk "may be used to describe a quantitative assessment of disease incidence or mortality in a defined population upon which policies such as annual screening mammography are built or it may describe a highly individual, subjective sense of danger, which might influence lifestyle 'choices'" (Aronowitz 2007, 5). In the context of treatment, risk may describe a quantitative assessment of a medical intervention in a defined population or it may describe a highly individualized sense of danger for a particular patient. In other words, risk assessment is never simply a value-free quantitative measure. It entails quantitative as well as qualitative determinations ranging from more objective measures of harm and benefit to more subjective ones (also see Lerner 2001, 11).

Given this, what are some general suggestions for navigating risk assessments in breast cancer care? According to Gigerenzer, who comments on navigating risk assessments in general, there are a number of steps that may be taken. First, we need to recognize "that almost all real-world events are uncertain and that we need to learn to deal with—rather than deny—this fact" (Gigerenzer 2002, 231).

The tools for taking this first step include considering "(1) real stories that illustrate the uncertainty inherent in everyday situations and (2) causal understanding of uncertainties and errors" (Gigerenzer 2002, 231). In the case of breast cancer, it would be helpful to have access to real stories and narratives that illustrate how actual breast cancer patients make the decisions they do about their care. It would also be helpful to have greater discussions of the uncertainties entailed in current breast cancer diagnosis, prognosis, and treatment.

Second, there is a need to overcome "both internal and external sources of ignorance about breast cancer. The goal is (1) to teach people how to use tools for estimating risks, including the uncertainty around these estimates, and (2) to make people aware of the forces aimed at preventing them from estimating risks" (Gigerenzer 2002, 233). In the case of breast cancer, it would be helpful if the American Cancer Society and the National Cancer Institute would include in their information pamphlets and on their websites information about how to understand and interpret basic risk assessments. Breast cancer specialists recommend to their patients that they read information coming out of such reputable institutions, but typically they do not help patients understand the medical information and how it applies to individual cases of breast cancer. There is usually not time in a clinical appointment for such discussions, largely because of the way insurers reimburse providers for the care they provide and because of the way health care professionals book patients back to back in short increments of time. And then there is this: there is little discussion in the clinical setting about the forces that prevent patients from considering risks in certain ways. Here one is reminded of Proctor's work (1995), which discloses some of the institutional and financial forces that frame available information and recommendations in certain ways in medicine.

Third, in navigating risk assessments in medicine, it is helpful to recognize that "[t]he idea that it is possible to communicate information in a 'pure' form is fiction" (Gigerenzer 2002, 239). In the case of breast cancer, it would be helpful to talk more about this in medicine. Consider the following example drawn from breast cancer medicine provided by Lerner concerning what constitutes "cure," "five-year cure," and "ten-year cure." For doctors, "cure" usually means that the cancer has not returned. For patients, this usually means the permanent disappearance of the disease (Lerner 2001, 1) This finding should not surprise us by this point of our investigation because, as we learned from the previous discussions, clinical information comes about through intersectional framings of the descriptions and explanations that are situated within contexts. Teasing out the forces that bring about clinical information is an important endeavor. Such an endeavor is important in breast cancer medicine in order to unpack how breast cancer specialists understand and treat breast cancer and how patients make the decisions that they do. A closer look at how patients make decisions will inevitably illustrate how contexts matter in how information is found, interpreted, and disseminated.

In the end, developing an appreciation of risk assessments is not just a matter of minimizing patient harm and maximizing patient welfare. It is a matter of clarifying and questioning our assumptions, priorities, and investments in breast cancer care. As Aronowitz reminds us, "we should not understand breast cancer risk ideas and terms as a merely logical or self-evident way of conceptualizing and communicating about danger, choice, cause, and responsibility" (2007, 6). He continues: "Modern risk discourse often reveals more about our present and past assumptions, priorities, and investments than it expresses new etiological, preventive, or therapeutic insights" (2007, 6). The call is to understand how breast cancer specialists and patients make the decisions they do and what they may need in order to make such decisions in light of fluid and evolving clinical information and the goals before them. Such a call is in keeping with an integrative or intersectional approach to understanding and treating breast cancer. The lesson here is that risk assessments are contextual and appeal to a host of considerations. One set of considerations involve the relation between appropriate risk assessments of breast cancer and how medicine structures access to breast cancer care, a topic that is considered in the next section.

ACCESS TO BREAST CANCER CARE AND JUSTICE

Access to breast cancer care in the United States has improved in the last forty years with the development of new diagnostics and treatments for breast cancer. It has improved as well with the passage of numerous laws and policies that address access to breast cancer care. Let's take a closer look at access to breast cancer care in the United States through the lens of the moral responsibility based on justice to secure it and the ethical challenges that emerge. Again, this is a broad topic, so I offer at best some general suggestions to think about.

The principle of justice guides our thinking about what is fair, equitable, and appropriate in light of what is due or owed to persons. The term "justice" comes from the Latin root *justus,* meaning "just" or "right." Justice is one of the oldest moral principles dating back to the Ancient Greek philosopher Aristotle (384–322 B.C.E.) (1941). As Aristotle says in the *Nicomachean Ethics,* "the just, then, is the lawful and the fair, the unjust the unlawful and the unfair" (1941, 1129a, 34–35). Common to all theories of justice is Aristotle's mandate to "treat equals equally" and "treat like cases alike." According to Beauchamp and Childress, "[t]his principle of justice . . . is 'formal' because it identifies no particular respects in which equals ought to be treated equally and provides no criteria for determining whether two or more individuals are in fact equals" (Beauchamp and Childress 2013, 250–251).

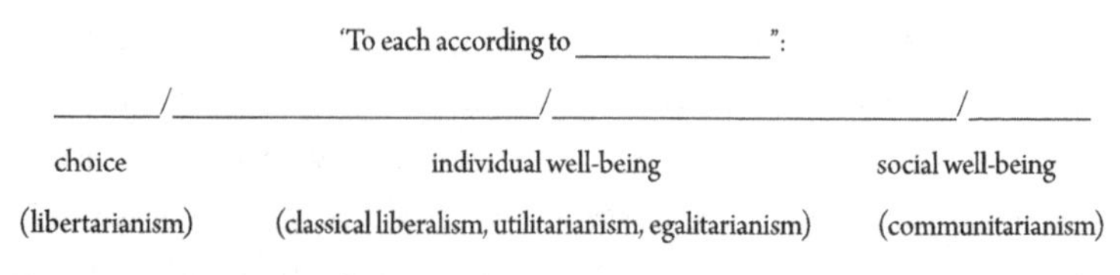

Figure 7.7 Spectrum of justice.

A variety of "material" or specific principles of justice have been proposed (Rawls 1971; Nozick 1974; Taylor 1979). Justice in health care typically concerns what constitutes proper distribution of resources. One way to conceive of this is in terms of the phrase, "To each according to _______." Framing the discussion of justice in terms of "to each according to _____," here are a few prominent ways to view justice (see Figure 7.7).

On one end of the spectrum is a libertarian view of justice. A libertarian view of justice holds that choice or autonomy is the most important appeal in a system of justice. As philosopher Robert Nozick says, "From each as they choose, to each as they are chosen" (1974, 160). In other words, a just system is one in which individuals can choose. On the other end of the spectrum is a communitarian view of justice. A communitarian view of justice holds that social or communal welfare is the most important appeal in a system of justice. A just system is one in which people's needs are satisfied. In the middle, so to speak, are classical liberal, utilitarian, and egalitarian views of justice. All hold that some balancing between individual autonomy and communal welfare is warranted. A just system is one that balances respect for choice with the satisfaction of needs. How this is worked out varies, as illustrated by classical liberal philosopher John Locke (1634–1704) (1924 [1689]), utilitarian philosopher John Stuart Mill (1806–1873) (1979 [1861]), and egalitarian philosopher John Rawls (1921–2002) (1971).

Putting differences aside (and recognizing that this is asking a lot), justice "refers to fair, equitable, and appropriate distribution of benefits and burdens determined by norms that structure the terms of social cooperation. Its scope includes policies that allot diverse benefits and burdens such as property, resources, taxation, privileges, and opportunities" (Beauchamp and Childress 2013, 250). It includes distribution according to equal share, need, effort, contribution, merit, and free-market exchanges (Beauchamp and Childress 2013, 251). The call for justice in breast cancer care highlights the moral imperative that "no persons should receive social benefits on the basis of undeserved advantageous properties (because no persons are responsible for having these properties) and that no persons should be denied social benefits on the basis of undeserved disadvantageous properties (because they also are not responsible for these properties)" (Beauchamp and Childress 2013, 248).

In the case of breast cancer, properties, such as breast cancer, are not chosen by individuals and thus do not provide grounds for morally accepted denial of or discrimination in health care. The call for justice in breast cancer care has been at the forefront of so many who practice in breast cancer medicine and those who have advocated for breast cancer patients. It has fueled efforts to lobby for legislation and policies that require access to health care services for breast cancer patients from diagnosis to survivorship care, including the Women's Health and Cancer Act of 1998, the Breast and Cervical Cancer Prevention and Treatment Act of 2000, the Affordable Care Act of 2010, and BreastScreen Singapore (see Chapter 1).

Nevertheless, there is work to be done. Given how medicine understands and treats breast cancer, there are a number of challenges with access to breast cancer care. Putting differences among material views of justice aside and focusing on the moral requirement found in every system of justice "to treat like cases alike," we confront the need to provide medical care to those who, through no fault of their own, express breast cancer. Given that there are means to diagnose breast cancer, and some means are widely available (such as the TNM staging system of classification for breast cancer), we can say that there is a moral imperative to ensure access to breast cancer screening and testing in developed as well as developing countries. Great strides have been made in developed countries, but there remains need in developing countries for better access to breast cancer care. As Danny Youlden and his colleagues report, although incidence rates of breast cancer are higher in developed countries, survival rates from breast cancer in Africa are far lower than in the United States, Australia, and Canada. They provide the following data:

> Almost 1.4 million women were diagnosed with breast cancer worldwide in 2008 and approximately 459,000 deaths were recorded. Incidence rates were much higher in more developed countries compared to less developed countries (71.7/100,000 and 29.3/100,000 respectively, adjusted to the World 2000 Standard Population) whereas the corresponding mortality rates were 17.1/100,000 and 11.8/100,000. Five-year relative survival estimates range from 12% in parts of Africa to almost 90% in the United States, Australia and Canada, with the differential linked to a combination of early detection, access to treatment services and cultural barriers. (Youlden et al. 2012, 1)

Given these data, like cases are certainly not treated alike.

In light of such data, it is difficult to critique the United States for its access to breast cancer care. But there are some notable studies that point us in the direction of further change and development. In the United States, a recent study shows that "[b]lack women with breast cancer . . . are on average about 40% more likely to die of the disease than white women with breast cancer" (Parker-Pope 2014, ¶1). Another study shows that those who are economically challenged tend not to request or know about reconstruction options after a mastectomy (Doheny 2014). Another study shows that there is a lack of standards for survivorship care in breast cancer medicine, including a lack of treatment of postsurgical pain and immobility, infertility care, and general follow-up care for women with breast cancer (Salz 2013). What these results show is room for improvement in the United States in access to breast cancer care.

In addition to the prior challenges to just access to breast cancer care, justice in breast cancer requires that we come to terms with what we know and what we do not know so that breast cancer patients are neither overdiagnosed and overtreated, nor underdiagnosed and undertreated. The overdiagnosis of breast cancer takes place when a patient is diagnosed with breast cancer when she does not have breast cancer. As previously mentioned, there has been significant debate in medicine about the overdiagnosis of ductal carcinoma in situ (DCIS). The view is that until medicine is able to determine which DCISs develop into late-stage cancer and which do not, clinicians will diagnosis and treat DCIS as breast cancer. As Dr. Clifford Hudis, chief of the Breast Cancer Medicine Service at Memorial Sloan-Kettering Cancer Center in New York City, says, "Even in the precancerous stage, called ductal carcinoma in situ (DCIS) when abnormal cells are confined to a milk duct, physicians almost always advise women to have a lumpectomy or mastectomy along with radiation, because about 20% of the 65,000 cases of DCIS found every year in the U.S. become invasive cancer" (Beck 2012, 1). Because of this, "[w]e do more than we need because we don't know how to do less" (Beck 2012, 1).

Overdiagnosis of breast cancer occurs in cases of DCIS where the DCIS does not develop into breast cancer. In such cases, DCIS is some aberration of the breast tissue, but it is not breast cancer (Beck 2012, 2014). Overtreatment of breast cancer takes place when a patient is treated for breast cancer when she does not have breast cancer. She is treated with a lumpectomy, mastectomy, chemotherapy, adjuvant therapy, and/or radiation when she does not have breast cancer (Beck 2012). Again, this typically occurs in cases in which clinicians are unsure about whether a condition will develop into a problem later for a patient, and so clinicians deem at this point that it is better to treat the condition than not to treat it, and patients willingly agree to or request such interventions.

Alternatively, the underdiagnosis and undertreatment of breast cancer are also issues. The underdiagnosis of breast cancer takes place when a patient is not diagnosed with breast cancer when she does in fact have breast cancer. Such occurred to me a few years before I was diagnosed with breast cancer. Underdiagnosis occurs in cases in which diagnostic screens and tests miss evidence for breast cancer. They miss it because the screens and tests—and clinical readers—are not 100% accurate. They miss it because our understanding of breast cancer is limited. Undertreatment of breast cancer takes place when a patient is not treated for breast cancer when she does have breast cancer. Although the problem of underdiagnosis and undertreatment of breast cancer is not at the forefront of current discussions in breast cancer care, it is nevertheless a legitimate focus of attention when discussing just access to breast cancer care.

The problem of the overdiagnosis and overtreatment, and underdiagnosis and undertreatment, of breast cancer results in false-negatives and false-positives, respectively. Such occurrence is an issue of justice because valuable resources are distributed to those who do not need them in the case of those who are overdiagnosed and overtreated with breast cancer. Alternatively, an issue of justice is raised when valuable resources are *not* distributed to those who need it. Inappropriate distribution of resources takes place when there are no clear medical guidelines for the diagnosis and treatment of breast cancer based on clinical evidence. In this way, the quest for knowledge of breast cancer and the need for better risk assessments in breast cancer care are matters of justice.

So how ought medicine to navigate the threshold between the overdiagnosis and overtreatment, and underdiagnosis and undertreatment, of breast cancer? I have four suggestions. A starting point is to recognize the uncertainty of breast cancer screening, tests, and treatments. Second, it will be helpful in breast cancer education in medical and nursing schools to devote time to discussing such uncertainty and how to understand risk assessments concerning breast cancer screening, testing, and treatment. Third, given the widespread use of mammograms, it will be important to revise and develop such screening technologies in order to reduce false-positive and false-negative results. If revision is not possible, what kind of alternative screens and tests might be possible? Fourth, it will be important for breast cancer specialists to discuss with patients the uncertainty of breast cancer screening, tests, and treatments and how such uncertainty plays out in decision making in breast cancer care. Just communication depends on finding the balance between truthful communication and the recognition of what is not known set within a framework of care (and not fear) in the health care professional–patient relationship. Just access to breast cancer care requires this balance as well.

The point is that just access to breast cancer care requires working with accurate risk assessments, where risk assessments are understood to be quantitative as well as qualitative assessments that are fraught with uncertainty. Finding the right balance between benefits and harms requires thinking through what constitutes the proper allocation of benefits and harms, which is not only an issue of nonmaleficence and beneficence, but of justice as well. Again, we find the need to think integratively in medicine, this time in the context of analyzing ethical issues in breast cancer care, a topic that is considered in the next section.

AN INTEGRATIVE ETHICAL APPROACH TO UNDERSTANDING BREAST CANCER

While the four principles of biomedical ethics are well received and widely used ethical principles in biomedical ethics, they were never meant by Beauchamp and Childress to operate alone. In Part III, Chapter 10 of their text, *Principles of Biomedical Ethics,* Beauchamp and Childress bring up a concept that few discuss when talking about a principle-based approach (or principlism) in biomedical ethics. This concept is one of "reflective equilibrium," which means "a way of bringing principles, judgments, and background theories into a state of equilibrium or harmony" (Beauchamp and Childress 2013, 404). What Beauchamp and Childress mean is that the principles were never meant to operate individually, never meant to be applied simplistically (e.g., simple "top-down" reasoning), and never meant to be arrived at simplistically (e.g., simple "bottom-up" reasoning). Consider the interconnectedness of biomedical ethical principles (see Figure 7.8).

In this framework, the principles serve not as isolated moral imperatives but as interacting moral guides as we seek "coherence of the overall set of beliefs that are accepted upon reflective examination" (Beauchamp and Childress 2013, 405).

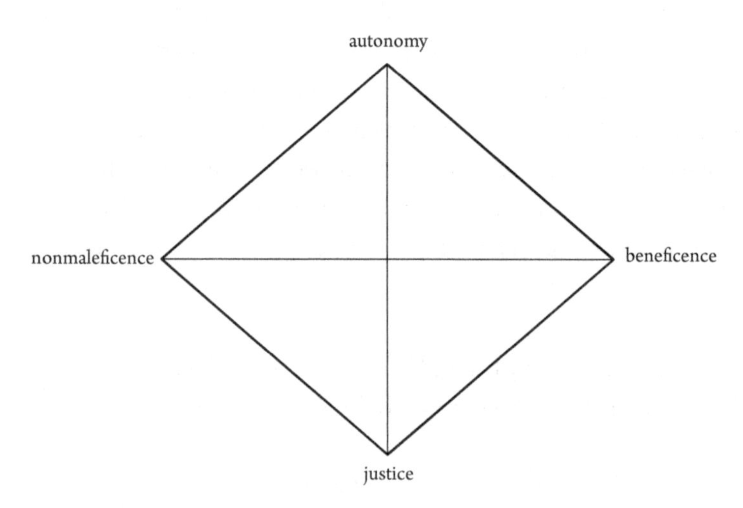

Figure 7.8 Interconnectedness of principles of biomedical ethics.

Reflective equilibrium "begins with a body of beliefs that are acceptable initially without argumentative support" (Beauchamp and Childress 2013, 405). These beliefs occur at all levels of moral thinking, from more concrete considerations through normative principles to metaethical conceptions (Beauchamp and Childress 2013, 405). As beliefs are developed and conflicts between and among beliefs emerge, individuals "must modify something in their viewpoint in order to achieve equilibrium" (Beauchamp and Childress 2013, 405). The purpose of reflective equilibrium "is to match, prune, and adjust considered judgments, their specifications, and other beliefs to render them coherent" (Beauchamp and Childress 2013, 405). Individuals "then test the resultant guides to action to see if they yield incoherent results" (Beauchamp and Childress 2013, 405). If it does, then the guides to action must be readjusted. What this means is that, in thinking through ethical issues in medicine and taking a stance, it will be important to do a good job specifying the ethical theories, principles, and values as they operate and intersect within specific contexts. By "specifying," Beauchamp and Childress mean "narrowing the scope of the norms" (2013, 17). Narrowing the scope of the norm involves "spelling out where, when, why, how, by what means, to whom, or by whom the action is to be done or avoided" (Richardson 1990, 289, cited in Beauchamp and Childress 2013, 17). Here reflective equilibrium involves a process of advancing, considering, and revising claims with regard to the problem under consideration.

Revising an example provided by Beauchamp and Childress (2013, 405), consider the case of the moral maxim "respect the breast cancer patient's autonomy." Contemporary biomedical ethicists place great emphasis on this maxim because patients are seen to be worthy of respect as individual decision makers. They seek to make this maxim as coherent as possible with other considered judgments about responsibilities in clinical practice, to patient's families, to hospitals, to the public, and so forth. Upon reflection, the moral maxim is not absolute but intersects with other moral maxims such as "do not harm the patient," "benefit the patient," "benefit the patient's family," "follow proper standards in clinical practice," "follow the rules of the hospital," "follow laws in society," and so on. While the moral maxim "respect the breast cancer patient's autonomy" is an acceptable starting premise, "[w]e are left with a range of options about how to specify this rule and check and balance it against other norms" (Beauchamp and Childress 2013, 405). For instance, the moral maxim to respect the breast cancer patient's autonomy is balanced with the moral maxim to not harm the patient, to benefit the patient, and to secure justice in health care. In the case of respecting a breast cancer patient's autonomy, attention is given to ensuring that the patient's welfare is promoted and justice is not compromised in health care. Compromises may have to take place, and when they do, it will be important to be clear on what is being compromised and in what ways in order to reach the best solution. In the end, "no matter which option we select, the coherence of our norms will always be a primary objective in the process of specification" (Beauchamp and Childress 2013, 405).

In this way, a principle-based approach in biomedical ethics can be seen to be integrative. It is integrative because it situates and defines moral theories, principles, and values contextually in the process of arriving at recommendations for what ought to be done in a certain situation. It is integrative because it calls us to match, prune, and adjust considered judgments, their specifications, and other beliefs to render them coherent and applicable to the situation. It requires that we see the moral principles of autonomy, nonmaleficence, beneficence, and justice not as solitary principles but as ones that define and situate each other within moral contexts of action in medicine.

Nonetheless, a principle-based approach in biomedical ethics is not without its critics. Chief among the critics come from those writing in feminist ethics. Feminist ethicists reject the traditional notion that ethics can be represented by a set of abstract principles and that morality of actions and policies can be assessed by reference to them. For feminists, ethics is part of an ongoing effort to uncover and eliminate sources of social inequality or oppression. Part of rooting out the sources is looking at the nature of ethics or morality itself, including the principles and maxims that it generates. For feminists, ethics is in many ways a product of men's power domination over women. Any emphasis on rights, for instance, shows preference toward men who were rights holders and property owners well before women. Feminist philosophers, such as Annette Baier (1992), Susan Sherwin (1992), and Rosemarie Tong (1997, 2014), are concerned about the ways medicine as a social institution generates, sustains, and uses power to marginalize persons. Unquestioned claims and assumptions about breast cancer, the medicalization of women's conditions in breast cancer care, and the lack of focus on the relation between environmental toxins and breast cancer all point toward areas in breast cancer medicine in which those in power have made decisions about health care that have affected breast cancer patients. An emphasis on the relational character of moral principles and on power relations in ethical discourse all point to important contributions feminists have made to discussions in biomedical ethics.

According to Rosemarie Tong, a principle-based bioethical approach to problems in medicine fails to be what she calls "eclectic," "autokoenomous," "positional," and "relational." By "eclectic," Tong means an approach that permits two or more politics "simultaneously, each of them serving as a corrective for each other's myopic tendencies" (Tong 1997, 93). By "autokoenomous," she means the ability to use two or more ontologies or views of reality "to better explain the paradox that consists in being a self whose individuality is necessarily constituted through the relationships with others" (Tong

1997, 94). By "positional," Tong means the ability to see knowledge as "situated in that it emerges from the roles and relationships an individual has" (Tong 1997, 95). By "relational," she means "steering a midcourse between an ethics of care and an ethics of power, each of which is relational in its own way" (Tong 1997, 96) and in need of one another. Tong calls us to see patients as decision makers within particular sociopolitical communities and our assessments of an ethical problem as lodged within local ontologies and epistemologies of difference.

Returning to the case study involving respecting a breast cancer patient's autonomy shared earlier, consider the case of the moral maxim "respect the breast cancer patient's autonomy." A feminist, intersectional approach to biomedical ethics seeks to make this maxim as coherent as possible with other considered judgments about responsibilities in clinical practice, to patient's families, to hospitals, to the public, and so forth. The focus of analysis is on the patient and her decisions in light of the context of her relationships. Upon reflection, the moral claim "respect the breast cancer patient's autonomy" must be interpreted to mean "respect the breast cancer patient's relationships," "reassess the care and power relations that influence the breast cancer patient's choices," and "promote the context in which healing can best take place." The moral claim "respect the breast cancer patient's autonomy" is to be seen as defined and situated within particular contexts in which the breast cancer patient lives and makes decisions. In the end, discussion of the importance of respecting a breast cancer patient's autonomy attends to the local ontologies, epistemologies, and ethical frameworks of those involved.

A lesson that I draw from Tong is to see discourses in biomedical ethics "as evolving and relational bodies of knowledge and values set within particular historical, cultural contexts" (Sassower and Cutter 2007, 131). In this way, the feedback loops discussed in Chapter 6 involving the diagnostic, therapeutic, and prognostic endeavors in breast cancer medicine take into consideration bioethical concerns. The resulting loops can be seen to be theoretically and practically fruitful in so far as they connect with the lives of those who practice medicine, those who seek the attention of health care professionals, and those who administer and oversee medicine. They will in the end be multifaceted, diverse, and dynamic as they take into consideration the complexity of the human condition and its differences. They will in the end be critical in so far as they allow us to step back and reflect on how we can improve breast cancer care.

CLOSING

Philosophically speaking, the question about what are the ethical implications of understanding breast cancer raises a host of issues, including informed consent, risk assessment, and access to breast cancer care. What we find is that, initially, there are adequate guidelines for informed consent in breast cancer care, adequate efforts to develop risk assessment measures in breast cancer care, and a decent level of access to breast cancer care in the United States, afforded largely by the attention given to the need for breast cancer care from advocacy groups and the passage of legislation requiring care from the time of screening through survivorship care. But things can be improved from an ethical standpoint. Upon reflection, the informed consent process would benefit from a more explicit discussion of uncertainty in breast cancer medicine and the ways breast cancer patients make decisions about their care. Risk assessments would benefit from a more personalized risk assessment approach. Access to breast cancer care can be improved by continued studies of the diverse forces that limit access to proper breast cancer care. Reflections on bioethical issues in breast cancer care are best framed integratively for the values, principles, and theories in moral philosophy are themselves situated and relational. These are not simply clinical and institutional suggestions for the practice of breast cancer care; they are moral ones, for medicine's long-standing commitment to patient autonomy, patient welfare, and social justice is at stake.

Chapter 8

Extended Musings

OPENING

When I began my journey as a breast cancer patient, I thought innocently that more would be known about breast cancer and more assurance could be made about its treatments. All I had to do would be to inform myself about breast cancer, ask the right questions, listen to the advice given to me by my breast cancer specialists, and follow protocols. Easy, right? Innocently, I thought, with all the funds going toward breast cancer research and care, thanks to the efforts of extraordinarily committed breast cancer advocates, patients, family members, researchers, and clinicians, the diagnosis and treatment of breast cancer would be somewhat uncluttered with confusing questions and difficult choices. Innocently, I thought, with all the efforts philosophers of medicine and bioethicists have put into developing the criteria for proper informed consent and risk assessment, the informed consent process and risk evaluation would be informative, meaningful, and empowering. Unbeknownst to me at the time, I was wrong. As a breast cancer patient, I have been struck with the lack of discussion about the ambiguities and uncertainties of breast cancer screening, testing, diagnosis, prognosis, treatment, and survivorship. I have been perplexed by the multitude of questions that have gone unanswered about my own case of breast cancer. Why did I have breast cancer, given that I have no family history or any other major risk factor for breast cancer? Why do so many women have breast cancer? What is going on with regard to women's breasts that lead each year to 226,000 cases in the United States and 1.7 million worldwide of newly diagnosed breast cancer? Why do 40,000 in the United States and 500,000 worldwide die from breast cancer? Why do we not understand how to prevent breast cancer? Why are we still treating breast cancer with such harsh treatments, such as a mastectomy and chemotherapy? Why do we not understand the different kinds of breast cancer? Why do we not understand the relation between breast cancer and other cancers? Why do we not have more targeted treatments for the different kinds of breast cancer? How ought we to design breast cancer care so that it best promotes patient autonomy, the minimization of patient harms, the maximization of patient benefits, and just access to breast cancer care? Here, once again, we are reminded of the statement by Aronowitz that has served as a major impetus in this investigation: "'What is breast cancer?' has been a recurrent, central, if often unarticulated, question just below the surface of so many controversies about cause, prevention, treatment, prognosis, and policy" (Aronowitz 2007, 7). It is also the question that "lies below the surface of many individual's difficult decisions" (2007, 7). So I ask myself in this final chapter, what lessons can be drawn from this philosophical reconstruction of how we understand and treat breast cancer today? What lessons can be drawn about how to navigate the uncertainties of breast cancer diagnosis, prognosis, and treatment? In this final chapter, I offer some concluding thoughts.

SEVEN LESSONS

There are seven lessons that I draw from this inquiry. First, I have learned that my breast cancer is located in the space of the physical world (and specifically the breast), the mind (and specifically the *idea* of breast cancer), and the history of views supported by a collective whole (which dates back many centuries). Today, breast cancer is distinguished in terms of kinds, kinds found in particular patients, brought together and generalized as categories that reflect shared properties and perceptions of breast cancer expressed in particular patients. As much as we may wish to reduce breast cancer to this or that kind, we cannot. Breast cancer is a complex, evolving medical phenomena expressed in individual patients. Nevertheless, the description of breast cancer is not completely open-ended. Breast cancer varies in terms of different kinds of cells, locations, sizes, and rates of growth. Such kinds, locations, sizes, and rates of growth can be assessed by mammography and, better yet, biopsy because mammograms provide limited information and interpreters of mammography information can miss things. Currently, medicine errs on the side of treating early-stage breast tumors because it is not able to determine which conditions develop into late-stage breast cancer and which do not.

Second, I have learned that I "got" breast cancer because my estrogen receptors on some of my breast cells were not functioning properly. Why did my estrogen receptors behave in this way? What caused my breast cancer and what is its prognosis? Why has my breast cancer been able so far to be controlled and other patients with a similar diagnosis have not had such good results? Who knows. What I have learned is that the cause of my breast cancer is at present less than certain, and clinicians do their best to prevent future occurrences and to treat the condition, given their limited understanding of breast cancer and the tools they have at their disposal. As a result, a healthy sense of skepticism is in order. A healthy sense of skepticism in breast cancer medicine encourages obtaining as much knowledge or information as possible, assessing the explanations that are forwarded, and translating population data into information that applies to individual patients. This speaks to the importance of translational work in breast cancer medicine, which interprets research results in ways that apply in the clinic. In this way, a healthy sense of skepticism peppered with the recognition of what we know about breast cancer helps navigate the jungle of information about why breast cancer occurs.

Third, I have learned that breast cancer diagnosis guides breast cancer treatment. Of course it does, one might say, but here I am suggesting that diagnosis guides treatment via a set of evaluative claims. Breast cancer diagnosis guides breast cancer treatment in terms of a range of values, some of which are quite shared and others that are less so. Shared values include the minimization of patient harms and the maximization of patient welfare. Less shared values include individual assessments of such harms and welfare considerations. The values here are an expression of those of individual patients, those of clinicians, those who pay for medical care, and those who regulate breast cancer medicine. The values that frame breast cancer are no more certain than the claims and assumptions that frame it, and so a healthy sense of skepticism is again in order. A healthy sense of skepticism peppered with the recognition of how we (dis)value breast cancer helps navigate the evaluations that are made in breast cancer care. It helps us understand why some breast cancer patients agree to certain kinds of interventions and others agree to other kinds. A healthy sense of skepticism reminds patients to continue to ask questions about their medical care. It reminds clinicians to look for individualized treatments for their breast cancer patients. A healthy sense of skepticism reminds me to take seriously my own evaluations of what is presented to me because my evaluations guide my medical treatment as I make decisions about my care along the way.

Fourth, I have learned that I have assumed a new social identity as a survivor, coward, warrior, and worrier. The range of possibilities is broad and relieves me from trying to be so strong all the time.

What is clear to me in terms of my new social identity is that I am not alone. I am a member of a club that I did not ask to join. The club involves having a condition recognized by breast cancer specialists, being evaluated by breast cancer specialists, and being considered a patient and one who is treated by breast cancer specialists. The club happens to receive a lot of attention these days—thank goodness for me and my fellow club members. But I am reminded that other clubs do not receive the same kind of attention, and this choice to attend to breast cancer and not other clinical conditions is a social endeavor itself. This choice involves decisions to fund certain endeavors in medicine and not others. It involves decisions to create and sustain disciplines, such as breast surgery and breast reconstruction, and not others. It involves decisions to classify and describe aspects of breast cancer and related clinical signs and symptoms as disease conditions, even when some of them might not currently be disease conditions. It involves decisions to advocate on behalf of breast cancer on a political level and not others. It involves decisions to advocate on behalf of groups of breast cancer patients and not others. Social forces play a central role in our understanding of breast cancer.

Fifth, I have learned that breast cancer is an integrative phenomenon. It is integrative because the descriptive, explanatory, evaluative, and social dimensions of my breast cancer are not separate and distinct. The dimensions define and situate each other by lending meaning and boundaries to our understanding and treatment of breast cancer. Descriptions of breast cancer define and situate the explanations of breast cancer, which define and situate evaluations of breast cancer, which define and situate socializations of breast cancer. Defining breast cancer in terms of Stage IIA, ER+ breast cancer sets up a certain set of explanations and evaluations of breast cancer that command a certain set of expectations in the social order of medicine to be treated in a certain way. Further, the relations are not always in this order. Evaluations frame descriptions and explanations, as seen in cases in which we treat breast cancer in order to prevent against harms, even when we do not fully understand the particular kind of breast cancer. Social forces frame evaluations, explanations, and descriptions as well. Rallying around the "cure" for breast cancer by vocal interest groups has led to significant efforts to describe, explain, and treat breast cancer. Breast cancer is an integrative notion that provides structure and significance to patient complaints and clinical reality.

Sixth, I have learned that, much to my dismay as a compliant patient and trained bioethicist, there were times when I did not give informed consent for my breast cancer care. I did not because sometimes I did not understand what I was signing up for. I did not understand because at those times I recognized the limits of my own clinical knowledge and clinical knowledge itself. I did not because at those times I was too overwhelmed with all the incoming information, the need to process it, and the need to make decisions in a timely fashion. Nevertheless, I did my best navigating the breast cancer information and risk assessments presented to me. Most times, I considered the information presented to me, understood it in terms of its consequences, spoke with others about it, obtained advice from others (including my clinicians, family members, and friends), made decisions based on it, authorized a treatment plan, and remained aware of the choices I made and their likely consequences. In some sense, this is the best I could do in a world of local, limited, uncertain, and changing knowledge. In a sense, this is what informed consent involves.

Further, I have learned that risk assessments are quantitative as well as qualitative evaluations. While I remain unclear about how exactly I made the decisions I did in light of the risks I perceived (or misperceived), I made decisions in light of a host of considerations and ones that went beyond sterile population data (which entail what I call "statistics without tears"). I considered some of the population data, sought advice about how to interpret the data, discussed with others how to think about the decisions before me, and made decisions the best I could. In a sense, this is what risk assessment involves.

Still further, I have realized that I am lucky and privileged in terms of my access to breast cancer care. I have employer-based insurance that covers most of my care and, when it does not pay, I pay for what I need—or what I am told that I need. In such cases when my insurer denied coverage, I challenged the insurance company because I feel empowered enough to challenge a large institution. But I know that this is not everyone's story. Health care disparities in breast cancer care exist within the United States as well as across the globe. There are those who do not know some of their options, do not have funds for their copays and medications, and do not have access to breast cancer services. Differences abound among breast cancer patients associated with race/ethnicity, gender/sexuality, class/economics, and age/ability. These differences affect susceptibility to breast cancer, the prognosis of those who express the disease, and responses to treatment. And I cannot help but reflect on whether just access to breast cancer care is compromised when the screenings and tests are not all that accurate, breast cancer risk assessments are cited so loosely, and treatment for breast cancer is not tailored to particular kinds of breast cancer. In this way, the over- and underdiagnosis, and over- and undertreatment, of breast cancer becomes a matter of social justice and should continue to command attention.

So, in answer to the central question in this inquiry, "What is breast cancer?," breast cancer is a family of diseases. Breast cancer is an evolving notion that provides structure and significance to patient narratives and clinical reality within social contexts. It is a relational process that describes, explains, evaluates, and socializes a clinical problem marked by dysfunctional breast cells programmed by biological pathways in the breast and perhaps elsewhere to multiply and survive in a patient. Breast cancer is a name we give to a problem in the breast that is considered harmful because, if left alone, it leads to pain, suffering, and perhaps death. It is a name we give to a disease that affects an anatomical region that commands great power, influence, and mystification in cultures. It is a treatment warrant that commands significant attention in our current culture because of our views of the breast, the harms that may be brought about, and our inability to win the "war on breast cancer." Breast cancer is a disease of civilization and the price we pay for living longer and within environments that support it. We can conclude with Aronowitz that "[b]reast cancer . . . has been and will continue to be transformed by what we believe about it and how we respond to it" (2007, 283).

There is one more lesson. Seventh, I have learned something about managing the uncertainty of breast cancer diagnoses, prognoses, and treatments. Breast cancer patients make the decisions they do about their breast cancer in the context of previously established frames of knowledge, predictions, and therapeutic actions. Coming to terms with such frames and how they influence the decisions that are made in breast cancer medicine constitutes the start to understanding the uncertainties that pervade breast cancer diagnosis, prognosis, and treatment. It is a start because such frames provide insight into the choice situations that women find themselves. For me, the framing of breast cancer that occurs in this inquiry leads me to this insight, one I draw from Engelhardt (1996, 221): one will not be able *simply* to discover, by appeal to factual issues alone, what diagnoses and treatments of breast cancer are indicated and what diagnoses and treatments are appropriate. One will have to consider the descriptions, explanations, evaluations, and social framings of such diagnostic and treatment decisions. Integral to such considerations will be appeals to particular views of reality, particular views of knowledge, particular hierarchies of values, and particular social expectations and forces. One will not be able simply to discover the answers; one creates them as well through appeal to a rich array of facts, explanations, values, and social appeals. Acknowledging such influences can shed light on the decisions one makes and the implications of one's decisions. Acknowledging such influences may lend some relief—relief from frantically searching for answers that do not exit. Acknowledging such influences highlights the responsibilities that come with clinical decision making, both for clinicians as well as patients.

Further, in managing clinical uncertainty, one must come to terms with what is *not* known about breast cancer. Here I am reminded of one of my favorite articles on the epistemology of ignorance authored by feminist philosopher Nancy Tuana (2006). Tuana tells us that if we are interested in knowing something, we are encouraged to reflect upon what we do not know and why. Paraphrasing her categories and applying them to our discussion of breast cancer, consider that there will be instances in which (1) we do not know what we do not know, (2) we do not know because we choose not to know, and (3) we do not know because others do not want us to know. As I began my journey as a breast cancer patient, (1) I was often frustrated with the sense that I did not know what was not known and I lacked an ability to even ask the questions. (2) I was often more than happy to put my head in the sand and not want to know anything more about breast cancer and my particular case of this clinical condition. Heck, I thought, I have a "right" not to know! Although this right has significant support in the biomedical ethical literature, I know now that my justification in this case arose from another motivation—to be left alone so that the problem would go away! (3) I was often frustrated with the lack of information on environmental contributions to breast cancer as well as the side effects of the array of options for pharmaceutical agents and radiation techniques presented to me. It just seemed and seems to me that there are backstories and lots more that we do not know about the nature, causes, and treatment of breast cancer. This is not to suggest that we can know everything about breast cancer, because we cannot. We cannot because humans are not omnipotent. Coming to terms with what we do not know helps recalibrate the scales of expectations about what we can know about breast cancer and ways to treat it. In this way, this inquiry is as much about managing expectations in the clinical setting as it is about figuring out what is breast cancer.

To aid with managing expectations, consider these seven additional pieces of advice:

1. Learn as much as possible about one's options and ask questions.
2. Reduce the timeline for decisions when making decisions about testing and treatment options.
3. Invest in keeping options open and keep one's goals and values at the forefront.
4. Take one risk or set of risks at a time and avoid unneeded risks.
5. Clarify what one knows and what one does not know throughout the process of making decisions.
6. Be flexible in decision making.
7. Accept the decisions one has made and minimize speculative guessing.

Remember: no discussion takes place in a vacuum. There is always a context and that context is important in decision making (Simon 1955; Bell, Raiffa, and Tversky 1988). There is always a process in the context, so take the time to engage in the process of shared decision making in the clinic (Braddock et al. 1997). And in the case of a majority of cases of breast cancer, there are two or more choices for care that are medically justified, thus making lots of room for patient preferences in the decision-making process (Lantz 2005).

And then there is this: after one has managed one's expectations and decided on a clinical course of screens, tests, biopsies, surgeries, and pharmaceutical interventions, "one will need to come to terms with the fact that the decisions one will make [or has made] will at times be wrong" (Engelhardt 1996, 220). In the case of past decisions, one will need to come to terms with the reality that one has made a series of decisions based on clinical information, statistical risks, personal desires, anxiety, fears, external forces, and an array of other considerations—and these have led one to where one is at present. Of course, there could have been more screens, tests, information, communication, and conversations. There could be more technology, treatments, and time. There could be less pressure, confusion, anxiety, and fear. But one chooses a clinical course of action in light of what is known and valued at the time, what is available, what is advised, what appears to work, and what seems to be acceptable. That's it. That's the best one can do. In some sense, the best one can do is determine what one knows, acknowledge what one does not know, act accordingly, and accept what one has decided within the context at hand. Then, and only then, can one move on.

In the case of my breast cancer, I have spent the years following my bilateral mastectomy, breast reconstruction, breast revisions, and pharmaceutical treatments coming to terms with what has occurred and what I agreed to occur based on an array of descriptions, explanations, evaluations, and social influences presented in this inquiry. Of course, things could have been different. But I chose a clinical course of action in light of my choice situations and did the best I could as a patient, wife, mother, daughter, grandmother, friend, colleague, bioethicist, and philosopher of medicine. I have acknowledged what I know and what I do not know, have acted accordingly, have begun to obtain a perspective of what has taken place, and am moving on. I am grateful for my excellent breast cancer care, for all that is known about breast cancer, for those who tirelessly advocate on behalf of breast cancer patients and their families, and for the opportunity to pen these reflections and be heard. While there is much more work to be done on breast cancer, there is much to celebrate about the strides that have been made in breast cancer medicine. Let us not forget to be grateful.

CLOSING

The foregoing is a philosophical reconstruction of breast cancer. It asks the question "What is breast cancer?" In answering it, I reconstruct the descriptive, explanatory, evaluative, and social dimensions of breast cancer and illustrate how the descriptive, explanatory, evaluative, and social dimensions of breast cancer are integrative. An integrative account of breast cancer carries ethical implications for how informed consent is secured, risks are assessed, and social justice is achieved in breast cancer care. As this inquiry shows, breast cancer is not an object simply to be discovered. It is a family of diseases. It is an evolving notion that provides structure and significance to patient narratives and clinical reality within social contexts. In the end, philosophy teaches us a number of lessons about how we understand and treat breast cancer and how to navigate the uncertainty of breast cancer diagnosis, prognosis, and treatment. Alternatively, breast cancer teaches us a number of philosophical lessons about our understanding of nature, knowledge, and what and how we value. My hope is that a fruitful dialogue between philosophy and medicine continues, with the particular hope that dialogues in breast cancer medicine continue to benefit patients. This is not just an academic wish; it is a personal one now that I have begun a "family history" of breast cancer.

GLOSSARY OF MEDICAL TERMS

The following definitions are taken from the National Cancer Institute (2014, 35–44).

advanced cancer: Cancer that has spread to other places in the body and usually cannot be cured or controlled with treatment.

aromatase inhibitor: A drug that prevents the formation of estradiol, a female hormone, by interfering with an aromatase enzyme. Aromatase inhibitors are used as a type of hormone therapy for postmenopausal women who have hormone-dependent breast cancer.

axilla: The underarm or armpit.

axillary dissection: Surgery to remove lymph nodes found in the armpit region. Also called axillary lymph node dissection.

axillary lymph node: A lymph node in the armpit region that drains lymph from the breast and nearby areas.

benign: Not cancer. Benign tumors may grow larger but do not spread to other parts of the body.

biopsy: The removal of cells or tissues for examination by a pathologist. The pathologist may study the tissue under a microscope or perform other tests on the cells or tissue.

blood vessel: A tube through which the blood circulates in the body. Blood vessels include a network of arteries, arterioles, capillaries, venules, and veins.

brachytherapy: A type of radiation therapy in which radioactive material sealed in needles, seeds, wires, or catheters is placed directly into or near a tumor. Also called implant radiation therapy, internal radiation therapy, and radiation brachytherapy.

breast-sparing surgery: An operation to remove the breast cancer but not the breast itself. Types of breast-sparing surgery include lumpectomy (removal of the lump), quadrantectomy (removal of one quarter, or quadrant, of the breast), and segmental mastectomy (removal of the cancer as well as some of the breast tissue around the tumor and the lining over the chest muscles below the tumor). Also called breast-conserving surgery.

cancer: A term for diseases in which abnormal cells divide without control and can invade nearby tissues. Cancer cells can also spread to other parts of the body through the blood and lymph systems.

carcinoma in situ: A group of abnormal cells that remain in the place where they first formed. They have not spread. These abnormal cells may become cancer and spread into nearby normal tissue. Also called stage 0 disease.

cell: The individual unit that makes up the tissues of the body. All living things are made up of one or more cells.

chemotherapy: Treatment with drugs that kill cancer cells.

clinical trial: A type of research study that tests how well new medical approaches work in people. These studies test new methods of screening, prevention, diagnosis, or treatment of a disease. Also called clinical study.

contrast material: A dye or other substance that helps show abnormal areas inside the body. It is given by injection into a vein, by enema, or by mouth. Contrast material may be used with x-rays, CT scans, MRI, or other imaging tests.

CT scan: A series of detailed pictures of areas inside the body taken from different angles. The pictures are created by a computer linked to an x-ray machine. Also called CAT scan, computed tomography scan, computerized axial tomography scan, and computerized tomography.

duct: In medicine, a tube or vessel of the body through which fluids pass.

ductal carcinoma: The most common type of breast cancer. It begins in the cells that line the milk ducts in the breast.

ductal carcinoma in situ: A noninvasive condition in which abnormal cells are found in the lining of a breast duct. The abnormal cells have not spread outside the duct to other tissues in the breast. In some cases, ductal carcinoma in situ may become invasive cancer and spread to other tissues, although it is not known at this time how to predict which lesions will become invasive. Also called DCIS and intraductal carcinoma.

early-stage breast cancer: Breast cancer that has not spread beyond the breast or the axillary lymph nodes. This includes ductal carcinoma in situ and stage I, stage IIA, stage IIB, and stage IIIA breast cancers.

estrogen: A type of hormone made by the body that helps develop and maintain female sex characteristics and the growth of long bones. Estrogens can also be made in the laboratory. They may be used as a type of birth control and to treat symptoms of menopause, menstrual disorders, osteoporosis, and other conditions.

external radiation therapy: A type of radiation therapy that uses a machine to aim high-energy rays at the cancer from outside of the body. Also called external-beam\radiation therapy.

fibrous: Containing or resembling fibers.

gland: An organ that makes one or more substances, such as hormones, digestive juices, sweat, tears, saliva, or milk.

HER2: A protein involved in normal cell growth. It is found on some types of cancer cells, including breast and ovarian. Cancer cells removed from the body may be tested for the presence of HER2/neu to help decide the best type of treatment. Also called c-erbB-2, human EGF receptor 2, and human epidermal growth factor receptor 2.

hormone receptor: A cell protein that binds a specific hormone. The hormone receptor may be on the surface of the cell or inside the cell. Many changes take place in a cell after a hormone binds to its receptor.

hormone therapy: Treatment that adds, blocks, or removes hormones. For certain conditions (such as diabetes or menopause), hormones are given to adjust low hormone levels. To slow or stop the growth of certain cancers (such as prostate and breast cancer), synthetic hormones or other drugs may be given to block the body's natural hormones. Sometimes surgery is needed to remove the gland that makes a certain hormone. Also called endocrine therapy, hormonal therapy, and hormone treatment.

inflammatory breast cancer: A type of breast cancer in which the breast looks red and swollen and feels warm. The skin of the breast may also show the pitted appearance called peau d'orange (like the skin of an orange). The redness and warmth occur because the cancer cells block the lymph vessels in the skin.

intravenous: Into or within a vein. Intravenous usually refers to a way of giving a drug or other substance through a needle or tube inserted into a vein. Also called IV.

leukemia: Cancer that starts in blood-forming tissue such as the bone marrow and causes large numbers of blood cells to be produced and enter the bloodstream.

LH-RH agonist: A drug that inhibits the secretion of sex hormones. In men, LH-RH agonist causes testosterone levels to fall. In women, LH-RH agonist causes the levels of estrogen and other sex hormones to fall. Also called luteinizing hormone-releasing hormone agonist.

lobe: A portion of an organ, such as the liver, lung, breast, thyroid, or brain.

lobular carcinoma: Cancer that begins in the lobules (the glands that make milk) of the breast.

lobular carcinoma in situ (LCIS): A condition in which abnormal cells are found only in the lobules. When cancer has spread from the lobules to surrounding tissues, it is invasive lobular carcinoma. LCIS does not become invasive lobular carcinoma very often, but having LCIS in one breast increases the risk of developing invasive cancer in either breast.

lobule: A small lobe or a subdivision of a lobe.

lumpectomy: Surgery to remove abnormal tissue or cancer from the breast and a small amount of normal tissue around it. It is a type of breast-sparing surgery.

lymph node: A rounded mass of lymphatic tissue that is surrounded by a capsule of connective tissue. Lymph nodes filter lymph (lymphatic fluid), and they store lymphocytes (white blood cells). They are located along lymphatic vessels. Also called lymph gland.

lymph vessel: A thin tube that carries lymph (lymphatic fluid) and white blood cells through the lymphatic system. Also called lymphatic vessel.

lymphedema: A condition in which excess fluid collects in tissue and causes swelling. It may occur in the arm or leg after lymph vessels or lymph nodes in the underarm or groin are removed or treated with radiation.

malignant: Cancerous. Malignant tumors can invade and destroy nearby tissue and spread to other parts of the body.

mammogram: An x-ray of the breast.

mastectomy: Surgery to remove the breast (or as much of the breast tissue as possible).

medical oncologist: A doctor who specializes in diagnosing and treating cancer using chemotherapy, targeted therapy, hormonal therapy, and biological therapy. A medical oncologist often is the main health care provider for someone who has cancer. A medical oncologist also gives supportive care and may coordinate treatment given by other specialists.

menopause: The time of life when a woman's ovaries stop working and menstrual periods stop. Natural menopause usually occurs around age 50. A woman is said to be in menopause when she hasn't had a period for 12 months in a row. Symptoms of menopause include hot flashes, mood swings, night sweats, vaginal dryness, trouble concentrating, and infertility.

menstrual period: The periodic discharge of blood and tissue from the uterus. From puberty until menopause, menstruation occurs about every 28 days, but it does not occur during pregnancy.

metastatic: Having to do with metastasis, which is the spread of cancer from one part of the body to another.

modified radical mastectomy: Surgery for breast cancer in which the breast, most or all of the lymph nodes under the arm, and the lining over the chest muscles are removed. Sometimes the surgeon also removes part of the chest wall muscles.

MRI: A procedure in which radio waves and a powerful magnet linked to a computer are used to create detailed pictures of areas inside the body. These pictures can show the difference between normal and diseased tissue. MRI makes better images of organs and soft tissue than other scanning techniques, such as computed tomography (CT) or x-ray. MRI is especially useful for imaging the brain, the spine, the soft tissue of joints, and the inside of bones. Also called magnetic resonance imaging.

oncology nurse: A nurse who specializes in treating and caring for people who have cancer.

organ: A part of the body that performs a specific function. For example, the heart is an organ.

ovary: One of a pair of female reproductive glands in which the ova, or eggs, are formed. The ovaries are located in the pelvis, one on each side of the uterus.

partial mastectomy: The removal of cancer as well as some of the breast tissue around the tumor and the lining over the chest muscles below the tumor. Usually some of the lymph nodes under the arm are also taken out. Also called segmental mastectomy.

PET scan: A procedure in which a small amount of radioactive glucose (sugar) is injected into a vein, and a scanner is used to make detailed, computerized pictures of areas inside the body where the glucose is used. Because cancer cells often use more glucose than normal cells, the pictures can be used to find cancer cells in the body. Also called positron emission tomography scan.

physical therapist: A health professional who teaches exercises and physical activities that help condition muscles and restore strength and movement.

plastic surgeon: A surgeon who specializes in reducing scarring or disfigurement that may occur as a result of accidents, birth defects, or treatment for diseases.

plastic surgery: An operation that restores or improves the appearance of body structures.

progesterone: A type of hormone made by the body that plays a role in the menstrual cycle and pregnancy. Progesterone can also be made in the laboratory. It may be used as a type of birth control and to treat menstrual disorders, infertility, symptoms of menopause, and other conditions.

radiation: Energy released in the form of particle or electromagnetic waves. Common sources of radiation include radon gas, cosmic rays from outer space, medical x-rays, and energy given off by a radioisotope (unstable form of a chemical element that releases radiation as it breaks down and becomes more stable).

radiation oncologist: A doctor who specializes in using radiation to treat cancer.

radiation therapy: The use of high-energy radiation from x-rays, gamma rays, neutrons, protons, and other sources to kill cancer cells and shrink tumors. Radiation may come from a machine outside the body (external-beam radiation therapy), or it may come from radioactive material placed in the body near cancer cells (internal radiation therapy). Systemic radiation therapy uses a radioactive substance, such as a radiolabeled monoclonal antibody, that travels in the blood to tissues throughout the body. Also called irradiation and radiotherapy.

radioactive: Giving off radiation.

reconstructive surgeon: A doctor who can surgically reshape or rebuild (reconstruct) a part of the body, such as a woman's breast after surgery for breast cancer.

registered dietitian: A health professional with special training in the use of diet and nutrition to keep the body healthy. A registered dietitian may help the medical team improve the nutritional health of a patient.

segmental mastectomy: The removal of cancer as well as some of the breast tissue around the tumor and the lining over the chest muscles below the tumor. Usually some of the lymph nodes under the arm are also taken out. Also called partial mastectomy.

sentinel lymph node biopsy: Removal and examination of the sentinel node(s) (the first lymph node(s) to which cancer cells are likely to spread from a primary tumor). To identify the sentinel lymph node(s), the surgeon injects a radioactive substance, blue dye, or both near the tumor. The surgeon then uses a scanner to find the sentinel lymph node(s) containing the radioactive substance or looks for the lymph node(s) stained with dye. The surgeon then removes the sentinel node(s) to check for the presence of cancer cells.

side effect: A problem that occurs when treatment affects healthy tissues or organs. Some common side effects of cancer treatment are fatigue, pain, nausea, vomiting, decreased blood cell counts, hair loss, and mouth sores.

social worker: A professional trained to talk with people and their families about emotional or physical needs, and to find them support services.

surgery: A procedure to remove or repair a part of the body or to find out whether disease is present. An operation.

tamoxifen: A drug used to treat certain types of breast cancer in women and men. It is also used to prevent breast cancer in women who have had ductal carcinoma in situ (abnormal cells in the ducts of the breast) and in women who are at a high risk of developing breast cancer. It blocks the effects of the hormone estrogen in the breast.

targeted therapy: A type of treatment that uses drugs or other substances, such as monoclonal antibodies, to identify and attack specific cancer cells. Targeted therapy may have fewer side effects than other types of cancer treatments.

tissue: A group or layer of cells that work together to perform a specific function.

total mastectomy: Removal of the breast. Also called simple mastectomy.

tumor: An abnormal mass of tissue that results when cells divide more than they should or do not die when they should. Tumors may be benign (not cancer) or malignant (cancer). Also called neoplasm.

x-ray: A type of high-energy radiation. In low doses, x-rays are used to diagnose diseases by making pictures of the inside of the body. In high doses, x-rays are used to treat cancer.

GLOSSARY OF PHILOSOPHICAL TERMS

acollaborative: The view that knowledge is not communal. Compare "collaborative."
aesthetic value: Value or assessment of beauty.
antirealism: The view that reality depends on human perception and thought.
apoliticalization: The process of not making a nonpolitical phenomenon a political one. Compare "politicalization."
autonomy: Ethical duty to respect self-determination. Autonomy involves two conditions, (1) liberty, or independence of controlling influences, and (2) agency, capacity for intentional action.
axiology: The philosophical study of value.
beneficence: Ethical duty to benefit another. Beneficence entails various kinds of duties, such as the protection and defense of the rights of others, the prevention of harm from occurring to others, the removal of conditions that will cause harm to others, help for persons with disabilities, and the rescue of persons who are in danger.
bioethics/biomedical ethics: The study of ethical issues in biomedicine. Bioethics/biomedical ethics emerged in the twentieth century in reaction to the atrocities that occurred in World War II.
collaborative: The view that knowledge is communal. Compare "acollaborative."
commodification: The process of making a nonsalable object a salable one.
contextualism: The view that claims depend on the context.
contributory causality: Correlative causal relation.
deduction: Inference from general to particular.
determinism: The view that every event has a cause.
empiricism: The view that knowledge is derived from sense perception.
epistemology: The philosophical study of knowledge.

ethics: The philosophical study of good and bad, and right and wrong, as these judgments have to do with the actions and character of individuals, families, communities, institutions, and society.

fact: An empirically verifiable statement that correlates with an object or state of affairs that exists.

feminist philosophy: The term for a host of positions that arise in response to the oppression of women and the need for accounts of reality, knowledge, and values that are responsive to the plights of marginalized persons.

functional value: Value or assessment of operation.

holism: The view that the whole is greater than the sum of its parts.

idealism: The view that reality is ideas.

induction: Inference from particular to general.

informed consent: In medicine, the process of obtaining permission from a patient before an intervention occurs. Involves disclosure of medical information, options, and consequences.

instrumental value: Value or assessment of means to end.

integrative: That which is brought together or incorporated into a whole. In medicine, refers to medicine that combines evidence-based medicine with alternative medicine.

intersectional: In philosophy and sociology, refers to a way of thinking that situates multiple perspectives with regard to each other.

justice: Ethical duty to allocate resources fairly.

libertarianism: The view that individuals should be free to exchange goods and services as they choose insofar as the participants in the arrangement agree.

maleficence: Ethical duty to benefit another.

materialism: The view that reality is composed of matter or that which is physical. See "physicalism."

medical thinking: The view for not making a nonmedical phenomenon a medical one. See "nonmedical thinking."

medicalization: The process of making something appropriate for medical intervention. Term is especially relevant when making a nonmedical phenomenon a medical one.

metaphysics: The philosophical study of that which is beyond the physical.

moral value: Value or worth of praiseworthiness or blameworthiness.

naturalism: The view that reality is composed of natural forces and entities.

necessary condition: A condition that must be satisfied in order for a statement to be true. Compare with "sufficient condition."

nonmaleficence: Ethical duty to minimize or avoid harm.

nonmedical thinking: The view for making a nonmedical phenomenon a medical one.

normativism: The view that norms or values operate in a claim of fact or understanding.

objectivity: The state of being independent of mind. Usually contrasts with subjectivity.
ontology: The philosophical study of being.
physicalism: The view that reality is physical or material. See "materialism."
politicalization: The process of making something a phenomenon by virtue of governmental or state actions. Term is especially relevant when making a nonpolitical phenomenon a political one. Compare "apoliticalization."
positivism: The view that knowledge is established through logic and mathematics.
probability: Extent to which some thing or event is likely to occur.
rationalism: The view that knowledge is derived from reason.
realism: The view that reality is independent of human perception or thought.
reason: A basis or cause of belief, action, facts, event, or sentiment.
reductionism: The view that a complex phenomenon is explainable in terms of its parts.
relativism: The view that claims are relative to cultural and social standards.
risk assessment: The measure of benefits over burdens or costs. A measure of the likelihood of adverse effects.
skepticism: The view that claims are to be doubted.
socialization: The process of making something fit communal or collective expectations. Term is especially relevant when making a social phenomenon a medical one.
spurious causality: Two events do not have direct causal connection.
subjectivity: The state of being dependent on the mind. Compare with "objectivity."
sufficient condition: A condition that must be satisfied in order for a statement to be true, but another condition can make the statement true. Compare with "necessary condition."
value: A sign of worth or assessment of comparison. In philosophy, a value can be nonmoral (as in the case of money) or moral (as in the case of respect for autonomy).
value neutralism: The view that norms or values do not operate in a claim or understanding. Compare "normativism."
value objectivism: The view that values or signs of worth are discovered and shared. Compare "value subjectivism."
value subjectivism: The view that values or signs of worth are personal. Compare "value objectivism."

REFERENCES

"The Affordable Care Act: How It Helps People with Cancer and Their Families." 2013. April 7. http://www.cancer.org/acs/groups/content/@editorial/documents/document/acspc-026864.pdf (accessed July 18, 2013).

American Board of Cosmetic Surgery. 2015. "Cosmetic Surgery vs. Plastic Surgery." http://www.americanboardcosmeticsurgery.org/patient-resources/cosmetic-vs-plastic-surgery (accessed July 3, 2015).

American Cancer Society. 2008. *Mastectomy: A Patient Guide*. Atlanta: American Cancer Society, Inc.

American Cancer Society. 2011. *For Women Facing Breast Cancer*. Atlanta: American Cancer Society, Inc. (No. 465200).

American Cancer Society. 2014a. "Americans with Disabilities Act: Information for People Facing Cancer." January 16. http://www.cancer.org./treatment/americancswith disabilitiesact/ (accessed February 27, 2014).

American Cancer Society. 2014b. "What Are the Risk Factors for Breast Cancer?" January 31. http://www.cancer.org/cancer/breastcancer/overviewguide/breast-cancer-overview-what-causes (accessed July 25, 2014).

American Cancer Society. 2015a. "Breast Cancer Screening Guidelines." http://www.cancer.org/healthy/informationforhealthcareprofessionals/acsguidelines/breastcancerscreeningguidelines/index (accessed June 2, 2015).

American Cancer Society. 2015b. "Breast Cancer Survival Rates—By Stage." http://www.cancer.org/cancer/breastcancer/detailedguide/breast-cancer-survival-by-stage (accessed June 23, 2015).

American Cancer Society. 2015c. "The History of Cancer." http://www.cancer.org/acs/groups/cid/documents/webcontrol/002048.pdf (accessed June 1, 2015).

American Cancer Society. 2015d. "Inflammatory Breast Cancer." http://www.cancer.org/cancer/breastcancer/moreinformation/inflammatorybreastcancer/inflammatory-breast-cancer-toc (accessed July 2, 2015).

American Cancer Society. 2015e. "What Are the Key Statistics for Breast Cancer?" http://www.cancer.org/cancer/breastcancer/detailed guide/breast-cancer (accessed April 21, 2015).

American Joint Committee on Cancer. 2010. *AJCC Cancer Staging Handbook.* New York: Springer.

American Joint Committee on Cancer. 2014. "About AJCC." http://cancerstaging.org/About/Pages/8th Edition.aspx (accessed March 1, 2014).

American Joint Committee on Cancer. 2017. *AJCC Cancer Staging Handbook.* New York: Springer.

Anderson, Elizabeth. 1995. "Feminist Epistemology: An Interpretation and Defense." *Hypatia* 10 (3): 50–84.

Anderson, Elizabeth. 2011. "Feminist Epistemology and Philosophy of Science." *The Stanford Encyclopedia of Philosophy* (Fall 2012 edition), Edward N. Zalta (ed.). http://plato.stanford.edu/archives/sum2012/entries/feminism-epistemology/

Annas, Julia, and Jonathan Barnes. 1985. *The Modes of Skepticism: Ancient Texts and Modern Interpretations.* Cambridge: Cambridge University Press.

Aristotle. 1941a. *Metaphysics.* In *The Basic Works of Aristotle,* edited and with introduction by Richard McKeon, 689–926. New York: Random House.

Aristotle. 1941b. *Nicomachean Ethics.* In *The Basic Works of Aristotle,* edited and with introduction by Richard McKeon, 935–1112. New York: Random House.

Aristotle. 1969. *Physics,* translated by H. G. Apostle. Bloomington: Indiana University Press.

Arizona Center for Integrative Medicine. 2009. "What Is IM?" http://integrativemedicine.arizona.edu/about (accessed December 15, 2009).

Aronowitz, Robert A. 2007. *Unnatural History: Breast Cancer and American Society.* Cambridge: Cambridge University Press.

Ayer, Alfred Jules. 1952. *Language, Truth, and Logic.* New York: Dover.

Baier, Annette. 1992. "Alternative Offerings to Aesclepius." *Medical Humanities Review* 6 (1): 9–19.

Bakalar, Marc. 2014. "Study Shows Third Gene as Indicator for Breast Cancer." *New York Times.* https://www.nytimes.com/2014/08/07/health/gene-indicator- breast-cancer-risk.html (accessed August 17, 2014).

Bartsky, Sandra Lee. 1988. "Foucault, Femininity, and the Modernization of Patriarchical Power." In *Femininity and Foucault: Reflections and Resistance,* edited by I. Diamond and L. Quinby, 61–86. Boston: Northwestern University Press.

Batt, Sharon. 1994. *Patient No More: The Politics of Breast Cancer.* Toronto: Gynergy Books.

Beauchamp, Tom L. 1982. *Philosophical Ethics: An Introduction to Moral Philosophy.* New York: McGraw-Hill.

Beauchamp, Tom L., and James F. Childress. 2013. *Principles of Biomedical Ethics*. 7th ed. New York: Oxford University Press.

Beck, Melinda. 2012. "Can There Be Too Much Breast Cancer Treatment?" *Wall Street Journal*, September 4, D1.

Beck, Melinda. 2014. "Some Cancer Experts See 'Overdiagnosis' Question Emphasis on Early Detection." *Wall Street Journal*, September 14. https://www.wsj.com/articles/some-cancer-experts-see-overdiagnosis-and-question-emphasis-on-early-detection-1410724838 (accessed September 15, 2014).

Beil, Laura. 2014. "Internal Signals." *Cure* 13 (1): 25–31.

Bell, D., H. Raiffa, and A. Tversky, ed. 1988. *Decision-making—Descriptive, Normative, and Prescriptive Interactions*. Cambridge: Cambridge University Press.

Berkeley, George. 1954. *Treatise Concerning the Principles of Human Knowledge*, edited by Colen M. Turbayne. New York: Macmillan.

Biological Sciences Curriculum Study. 1992. *Mapping and Sequencing the Human Genome: Science, Ethics, and Public Policy*. Colorado Springs: BSCS and American Medical Association.

Bliss, Michael. 1999. *William Osler: A Life in Medicine*. Oxford: Oxford University Press.

Boorse, Christopher. 1977. "Health as a Theoretical Concept." *Philosophy of Science* 44: 542–573.

Boorse, Christopher. 1997. "A Rebuttal on Health." In *What Is Disease?*, edited by J. M. Humber and R. F. Almeder, 3–171. New York: Humana Press.

Bowman, Deborah et al. 2012. *Informed Consent: A Primer for Clinical Practice*. Cambridge: Cambridge University Press.

Boyles, Salynn. 2012. "Do False-Positive Mammograms Predict Cancer Risk?" *WebMD Health News*, April 5. http://www.webmd.com/women/news/201204005 (accessed August 7, 2014).

Braddock, Clarence H., Stephan D. Fihn, Wendy Levinson, Albert R. Jonsen, and Robert A. Pearlman. 1997. "How Doctors and Patients Discuss Routine Clinical Decisions." *Journal of General Internal Medicine* 12: 339–345.

"Breast Cancer in Men." 2014. http: https://ww5.komen.org/BreastCancer/BreastCancerinMen.html (accessed March 6, 2014).

Breastcancer.org. 2015. http://www.breastcancer.org/ (accessed June 4, 2015).

Brinton, Roberta Diaz et al. 2008. "Progesterone Receptors: Form and Function in the Brain." *Front Neuroendocrinology* 29 (2): 313–39.

Brown, Zora K., and Harold Freeman, with Elizabeth Platt. 2007. *100 Questions and Answers About Breast Cancer*. New York: Jones and Bartlett Publishers.

Burger, Edward B., and Michael Starbird. 2005. *Coincidences, Chaos, and All That Jazz: Making Light of Weighty Ideas*. New York: W.W. Norton.

Canadian Cancer Society. 2007. "Breast Cancer Screening in Your 40s." http://www.cancer.ca/en/prevention-and-screening/early-detection-and-screening/screening/screening-for-breast-cancer/breast-cancer-screening-in-your-40s/?region=ab (accessed June 23, 2015).

"Cancer: The Generic Impact." 2008. *BioPortfolio Ltd*, May 16.

Caplan, A.L., H.T. Engelhardt, Jr., and J.J. McCartney (eds.). 1981. *Concepts of Health and Disease*. Boston: Addison-Wesley.

Carnap, Rudolf. 1966. *Philosophical Foundations of Physics*. New York: Basic Books.

Centers for Disease Control and Prevention. 2013. National Breast and Cervical Early Detection Program. http://www.cdc.gov/cancer/nbccedp/legislation August 2013 (accessed August 22, 2013).

Cheung, Y. F. 2017. "Medical Humanities in Paediatric Clinical Practice." *Hong Kong Journal of Paediatrics* 22: 70–71.

Christakis, Nicholas. 2011. "What Scientific Concept Would Improve Everybody's Cognitive Toolkit?" *Edge*. http://edge.org/q2011/q11_6.html (accessed September 22, 2014).

Code, Lorraine. 2003 (1991). "Is the Sex of the Knower Epistemologically Significant?" In *The Theory of Knowledge*, 6th ed., edited by Louis P. Pojman and James Feiser, 559–571. Belmont, CA: Wadsworth.

Collins, Francis. 2010. *The Language of Life: DNA and the Revolution in Personalized Medicine*. New York: HarperCollins Publishers.

Conrad, Peter. 2007. *The Medicalization of Society: On the Transformation of Human Conditions into Treatable Disorders*. Baltimore: Johns Hopkins University Press.

Conrad, Peter, and J. Schneider. 1980. "Looking at Levels of Medicalization: A Comment on Strong's Critique of the Thesis of Medical Imperialism." *Social Science and Medicine* 14A: 75–79.

Consigny, Scott. 2001. *Gorgias: Sophist and Artist*. Columbia: University of South Carolina Press.

Cornell, Drucilla. 2004. "Gender in America." In *Keywords: Gender—For a Different Kind of Globalization*, edited by N. Tazi, 33–54. New York: Other Press.

Crenshaw, Kimberlé. 1989. "Demarginalizing the Intersection of Race and Sex: A Black Feminist Critique of Antidiscrimination Doctrine, Feminist Theory, and Antiracist Politics." *University of Chicago Legal Forum* 14: 538–554.

Crenshaw, Kimberlé Williams. 1991. "Mapping the Margins: Intersectionality, Identity Politics, and Violence Against Women of Color." *Stanford Law Review* 43: 1241–1299.

Curigliano, G., G. Spitaleri, M. Dettori, M. Locatelli, E. Scarano, and A. Goldhirsch. 2007. "Immunology and Breast Cancer: Therapeutic Cancer Vaccines." *Breast* (December 16, Suppl. 2): S20–S26.

Cutter, Mary Ann G. 1992. "Value Presuppositions in Diagnosis: A Case Study of Cervical Cancer." In *The Ethics of Diagnosis*, edited by Jose Luis Peset et al., 147–154. Dordrecht: Kluwer Academic.

Cutter, Mary Ann G. (ed.). 2002. "Molecular Genetics and the Transformation of Medicine." *Journal of Medicine and Philosophy* 27 (3): 251–256.

Cutter, Mary Ann G. 2003. *Reframing Disease Contextually*. New York: Springer.

Cutter, Mary Ann G. 2012. *The Ethics of Gender-Specific Disease*. New York: Routledge.

Davis, Kathy. 1993. *Reshaping the Female Body: The Dilemma of Cosmetic Surgery.* New York: Routledge.

Davis, Kathy. 2008. "Intersectionality as Buzzword: A Sociology of Science Perspective on What Makes a Feminist Theory Successful." *Feminist Theory* 9: 67–85.

Dawkins, Richard. 1976. *The Selfish Gene.* New York: Oxford University Press.

DeGregorio, Michael W., and Valerie J. Wiebe. 1996. *Tamoxifen and Breast Cancer.* New Haven, CT: Yale University Press.

Descartes, René. 1911. *Discourse on Method.* In *Philosophical Works of Descartes,* translated by Elizabeth S. Haldane and G. R. T. Ross, I, 85. Cambridge: Cambridge University Press.

"Diagnosing Breast Cancer." 2014. http://www.breastcancer.org/symptoms/types/dcis/diagnosis (accessed August 25, 2014).

Doheny, Kathleen. 2014. "Many Women Who Have Mastectomy Don't Get Breast Reconstruction." August 20. https://consumer.healthday.com/cancer-information-5/breast-cancer-news-94/many-women-who-have-mastectomy-don-t-get-breast-reconstruction-study-690972.html (accessed September 17, 2014).

Ehrenreich, Barbara. 2001. "Welcome to Cancerland: A Mammogram Leads to a Culture of Pink Kitsch." *Harpers* (November): 43–53.

Engel, George. 1981. "The Need for as New Medical Model: A Challenge for Biomedicine." In *Concepts of Health and Disease: Interdisciplinary Perspectives,* edited by A.L. Caplan, et al., 589–607. Boston: Addison-Wesley.

Engelhardt, H. Tristram, Jr. 1981 [1975]. "The Concepts of Health and Disease." In *Concepts of Health and Disease: Interdisciplinary Perspectives,* edited by A. L. Caplan et al., 31–46. Boston: Addison-Wesley.

Engelhardt, H. Tristram, Jr. 1984. "Clinical Problems and the Concept of Disease." In *Health, Disease, and Causal Explanations in Medicine,* edited by L. Nordenfelt and B. I. B. Lindahl, 27–41. Dordrecht: D. Reidel.

Engelhardt, H. Tristram, Jr. 1986. "From Philosophy *and* Medicine to Philosophy *of* Medicine." *Journal of Medicine and Philosophy* 11: 3–8.

Engelhardt, H. Tristram, Jr. 1996. *Foundations of Bioethics.* 2nd ed. New York: Oxford University Press.

Engelhardt, H. Tristram, Jr. 2000. "The Philosophy of Medicine and Bioethics: An Introduction to the Framing of a Field." In *The Philosophy of Medicine: Framing the Field,* edited by H. T. Engelhardt, Jr., 1–15. Dordrecht: Kluwer Academic.

Engelhardt, H. Tristram, Jr., and Edmund L. Erde. 1980. "Philosophy of Medicine." In *A Guide to the Culture of Science, Medicine, and Technology,* edited by P. T. Durbin, 364–461. New York: Free Press.

Engelhardt, H. Tristram, Jr. and Kevin Wm. Wildes. 2003. "Health and Disease: Philosophical Perspectives." In *Encyclopedia of Bioethics,* 3rd ed., edited by S. G. Post, 1101–1106. New York: Macmillian Reference Books.

Ericksen, Julia. 2008. *Taking Charge of Breast Cancer.* Oakland: University of California Press.

Faden, Ruth, and Tom L. Beauchamp. 1986. *A History and Theory of Informed Consent*. New York: Oxford University Press.

Fleck, Ludwik. 1979 (1935). *Genesis and Development of a Scientific Fact*, edited by T.J. Trenn and R. K. Merton and translated by F. Bradley and T. J. Trenn. Chicago: University of Chicago Press.

Foucault, Michel. 1973 [1963]. *Birth of the Clinic: An Archeology of Medical Perception*, translated by A.M. Sheridan Smith. New York: Pantheon Books.

Fuller, Steve. 1988. *Social Epistemology*. Bloomington: Indiana University Press.

Gardell, Stephen J. 2014. E-mail communication, March 6.

Gigerenzer, Gerd. 2002. *Calculated Risks: How to Know When Numbers Deceive You*. New York: Simon and Schuster.

Gillham, Nicholas Wright. 2011. *Genes, Chromosomes, and Disease: From Simple Traits, to Complex Traits, to Personalized Medicine*. New York: Pearson Educational.

Gottlieb, Scott. 2000. "Lumpectomy as Good as Mastectomy for Tumors up to 5 cm." *British Medical Journal* (July 29), 261.

Gøtzsche, Peter C. 2007. *Rational Diagnosis and Treatment: Evidence-Based Clinical Decision-Making*. 4th ed. New York: John Wiley and Sons.

Grady, Denise. 2013. "Breast Cancer Drugs Urged for Healthy High-Risk Women." *New York Times*, April 15. http://www.nytimes.com/2013/04/16/health/breast-cancer-drugs-urged-for-healthy-high-risk-women.html?_r=0 (accessed May 11, 2015).

Grasswisk, Heidi. 2013. "Feminist Social Epistemology." *Stanford Encyclopedia of Philosophy*. http://plato.stanford.edu/entries/feminist-social-epistemology/ (accessed June 8, 2015).

Greenhalgh, Trisha. 2010. *How to Read a Paper: The Basics of Evidence-Based Medicine*. 4th ed. New York: John Wiley and Sons.

Groopman, Jerome. 2007. *How Doctors Think*. Boston: Houghton Mifflin.

Health Promotion Board. 2015. "BreastScreen Singapore." http://hpb.gov.sg/HOPPortal/health-article/3324 (accessed May 23, 2015).

Hippocrates. 1923. *Hippocrates and the Fragments of Heraclitus*. 4 vols. Translated by W.H. S. Jones. Cambridge, MA: Harvard University Press.

Hobbes, Thomas. 1950. *Leviathan*, edited by A.D. Lindsay, ch. 46. New York: Dutton.

Hume, David. 1894. *Enquiry Concerning Human Understanding*, edited by L.S. Selby-Bigge. Oxford: Clarendon.

Jacobsen, Nora. 1998. "The Socially Constructed Breast: Breast Implants and the Medical Construction of Need." *American Journal of Public Health* 88: 1254–1261.

Jacobsen, Nora. 2000. *Cleavage: Technology, Controversy, and the Ironies of the Man-Made Breast*. New Bruswick, NJ: Rutgers University Press.

Jasanoff, Sheila. 2004. *States of Knowledge: The Co-Production of Science and Social Order*. New York: Routledge.

Jolie, Angelina. 2013. "My Medical Choice." *The New York Times*, May 14. http://www.nytimes.com/2013/05/14/opinion/my-medical-choice.html (accessed May 30, 2014).

Jolie, Angelina. 2015. "Diary of a Surgery." *The New York Times*, March 24. http:www.nytimes.com/2015/03/24/opinion/Angelina-jolie-pitt-diary-of-a-surgeryu.html (accessed April 19, 2015).

Kant. 1985 [1785]. *Foundations of the Metaphysics of Morals*, translated by L.W. Beck. New York: Macmillan.

Kasper, Anne S., and Susan J. Ferguson. 2000. "Living with Breast Cancer." In *Breast Cancer: Society Shapes an Epidemic*, edited by A. S. Kasper and S. J. Ferguson, 1–22. New York: St. Martin's Press.

Khushf, George. 1997. "Why Bioethics Needs the Philosophy of Medicine: Some Implications of Reflection on Concepts of Health and Disease." *Theoretical Medicine and Bioethics* 18: 145–163.

Kligler, Benjamin, and Roberta Lee. 2004. *Integrative Medicine: Principles for Practice*. New York: McGraw-Hill.

Koopsen, Cyndie, and Caroline Young. 2009. *Integrative Health: A Holistic Approach for Health Professionals*. Boston: Jones and Bartlett.

Kuhn, Thomas. 1970. *The Structure of Scientific Revolutions*. 2nd ed. Chicago: University of Chicago Press.

Kushner, Rose. 1975. *Breast Cancer: A Personal and an Investigative Report*. New York: Harcourt Brace Jovanovich.

Laertius, Diogenes. 1925. *Lives of Eminent Philosophers*, translated by R. D. Hicks. Cambridge, MA: Harvard University Press.

La Mettrie, Julien Offray de. 1912 [1748]. *Man as a Machine*, translated by Gertrude C. Bussey et al. La Salle, IL: Open Court.

Lantz, Paula M., Nancy K Janz, Angela Fagerlin, Kendra Schwatz, Lihua Liu, Indu Lakhami, Barbara Salem, and Steven J. Katz. 2005. "Satisfaction with Surgery Outcomes and the Decision Process in a Population-Based Sample of Women with Breast Cancer." *Health Services Research* 40 (3): 745–767.

Lautner, Meeghan, Heather Lin, Yu Shin, Catherine Parker, Henry Kuerer, Simona Shaitelman, Gildy Babiera, and Isabelle Bedrosian. 2015. "Disparities in the Use of Breast-Conserving Therapy Among Patients with Early-Stage Breast Cancer." *JAMA Surgery* (June 17). doi:10.1001/jamasurg.2015.1102 (accessed July 3, 2015).

Lee, Jaimy. 2013. "Quest Diagnostics to Offer Genetic Tests for Breast Cancer Risk." *Modern Healthcare* (October 16). http://www.modernhealthcare.com/article/20131016/NEWS (accessed June 22, 2015).

Lerner, Barron H. 2001. *The Breast Cancer Wars: Hope, Fear, and the Pursuit of a Cure in the Twentieth Century America*. New York: Oxford.

Lerner, Barron H. 2006. "Power, Gender, and Pizzaz: The Early Years of Breast Cancer Activism." In *The Voice of Breast Cancer in Medicine and Bioethics*, edited by M. Rawlinson and S. Lundeed, 21–30. New York: Springer.

Locke, John. 1924 (1689). *An Essay Concerning Human Understanding*, edited by A.S. Pringle-Pattison. London: Oxford University Press.

Longino, Helen. 2002. *The Fate of Knowledge*. Princeton, NJ: Princeton University Press.

Lorde, Audre. 1980. *The Cancer Journals*. San Francisco: Spinsters/Aunt Lute.

Love, Susan M. 2010. *Dr. Susan Love's Breast Book*. 5th ed. Philadelphia: A Merloyd Lawrence Book.

Marcum, James A. 2008. *An Introductory Philosophy of Medicine: Humanizing Modern Medicine*. New York: Springer.

Marcum, James A. 2009. *The Conceptual Foundation of Systems Biology: An Introduction*. New York: Nova Science.

MedLine Plus. 2014. "Mammography Cuts Breast Cancer Deaths by 28 Percent: Study." June 18. http://www.nlm.nih.gov/medlineplus/news/fullstory_146868.html (accessed August 26, 2014).

Mill, John Stuart. 1979 (1861). *Utilitarianism*, edited with an introduction by G. Sher. Indianapolis: Hackett.

Morgagni, Giovanni. 1981 (1761). "The Author's Preface." In *Concepts of Health and Disease: Interdisciplinary Perspectives*, edited by A. L. Caplan et al., 157–165. Boston: Addison-Wesley.

Morgan, Kathryn Pauly. 1991. "Women and the Knife: Cosmetic Surgery and the Colonization of Women's Bodies." *Hypatia* 63: 25–53.

Mukherjee, Sidhartha. 2010. *The Emperor of All Maladies: A Biography of Cancer*. New York: Scribner.

Mullings, Leith, and Amy J. Schulz. 2006. "Intersectionality and Health." In *Gender, Race, Class, and Health*, edited by A.J. Schulz and L. Mullings, 3–17. Baltimore: John Wiley and Sons.

Murphy, Edmond 1976. *The Logic of Medicine*. Baltimore: Johns Hopkins University Press.

National Cancer Institute. 2014a. "Breast Cancer." January 16, p. 1. http://www.cancer.gov/cancertopics/types/breast

National Cancer Institute. 2014b. "Cancer Research Funding." http://www.cancer.gov/cancertopics/factsheet/NCI/research-funding (accessed August 18, 2014).

National Cancer Institute. 2014c. "Glossary of Medical Terms." http://www.cancer.gov/publications/patient-education/WYNTK_breast.pdf (accessed March 6, 2014).

National Cancer Institute. 2014d. "What You Need to Know About Breast Cancer." National Institutes of Health. http://www.cancer.gov/cancertopics/wyntk/breast, posted April 2012 (accessed November 13, 2014).

National Cancer Institute. 2015a. "BRCA1 and BRCA2: Cancer Risk and Genetic Testing." http://www.cancer.gov.cancertopics/causes-prevention/genetics/brca-fact-sheet (accessed April 1, 2015).

National Cancer Institute. 2015b. "Cancer Prevalence and the Cost of Care Projections." http://cost.projections.cancer.gov/expenditure.html (accessed January 6, 2015).

National Cancer Institute. 2015c. "Genetics of Breast and Gynecological Cancers." http://www.cancer.gov/cancertopics/pdq/genetics/breast-and-ovarian (accessed April 20, 2015).

National Cancer Institute. 2015d. "Hormone Therapy for Breast Cancer." http://www.cancer.gov/types/breast-hormone-therapy-fact-sheet (accessed June 20, 2015).

National Cancer Institute. 2015e. "International Cancer Screening Network." http://appliedresearch.cancer.gov/icsn/breast/screening.html (accessed June 8, 2015).

National Cancer Institute. 2015f. "Surveillance, Epidemiology, and End Results Program." http://www.seer.cancer.gov/statsfacts/hyml/breast.html (accessed April 5, 2015).

National Institute of Environmental Health Sciences. 2012. "Breast Cancer Risk and Environmental Factors." http://www.niehs.nih.gov/health/assets/docs_a_e/environmental-facts-and-breast-cancer-risk-508.pdf (accessed April 15, 2015).

Nelson, Lynn Hankinson. 1990. *Who Knows: From Quine to Feminist Empiricism*, Philadelphia: Temple University Press.

"Novel Immunotherapy Vaccine Decreases Recurrence in HER2 Positive Breast Cancer Patients." 2014. M. D. Anderson Cancer Center Newsroom, September 4. http://www.mdanderson.org/newsroom/news-releases/2014/vaccine-decreases-reoccurence-her2-breast-cancer.html (accessed September 24, 2014).

Nozick, Robert. 1974. *Anarchy, State, and Utopia*. New York: Basic Books.

Olson, James S. 2002. *Bathsheba's Breast: Women, Cancer, and History*. Baltimore: The Johns Hopkins University Press.

Pagel, Walter. 1958. *Paracelsus: An Introduction to Philosophical Medicine in the Era of the Renaissance*. Berlin: S. Karger.

Parker-Pope, Tara. 2014. "The Breast Cancer Racial Gap." *New York Times*, March 3. https://well.blogs.nytimes.com/2014/03/03/the-breast-cancer-racial-gap/ (accessed August 7, 2014).

Parsons, Talcott. 1951. *The Social System*. New York: The Free Press.

Pascal, Blaise. 1910 (1670). *Pensées*. Translated by W. F. Trotter. London: Dent.

Patient Resource. 2011. *Patient Resource Breast Cancer Guide*. Overland Park, KS: PRP Patient Resource Publishing. http://www.patientresource.net

Pellegrino, Edmund D. 1978. "Philosophy of Medicine: Towards a Definition." *Journal of Medicine and Philosophy* 11: 9–16.

Pellegrino, Edmund D. 2004. "Forward: Renewing Medicine's Basic Concept." In *Health, Disease, and Illness: Concepts in Medicine*, edited by A. L. Caplan et al., xi–xiv. Washington, D.C.: Georgetown University Press.

Pellegrino, Edmund D., and David Thomasma. 1981. *A Philosophical Basis of Medical Practice*. New York: Oxford University Press.

"Plastic Surgeon Job Description." 2015. *Healthcare Salaries*. http://www.healthcare-salaries.com/physicians/plastic-surgeon-salary (accessed June 4, 2015).

Plato. 1961. *Timaeus*, translated by Benjamin Jewett. In *Plato: The Collected Dialogues*, edited by Edith Hamilton and Huntington Cairns. New York: Pantheon Books.

Plutynski, Anya. 2012. "Ethical Issues in Cancer Screening and Prevention." *The Journal of Medicine and Philosophy* 37 (3): 310–323.

Pojman, Louis P. and James Feiser. 2011. *Ethical Theory*. 6th ed. Belmont, CA: Wadsworth.

Proctor, Robert N. 1995. *Cancer Wars: How Politics Shapes What We Know and Don't Know About Cancer.* New York: Basic Books.

Production of ICD-11: The Revision Process. 2007. http://www.who.int/classifications/icd/ICDRevision.pdf (accessed October 4, 2010).

"Radiologists Interpret Mammograms Differently." 2007. December 11. http://www.breastcancer.org/research-news/ (accessed August 7, 2014).

Rawls, John. 1971. *A Theory of Justice*. New York: Cambridge University Press.

Reale, Giovanni. 1987. *A History of Ancient Philosophy. Vol. I. From the Origins to Socrates.* Translated by J. R. Catan. New York: State University of New York Press.

Reznek, Lawrie. 1987. *The Nature of Disease*. London: Routledge and Keegan Paul.

Richardson, Henry S. 1990. "Specifying, Balancing, and Interpreting Bioethics Principles." *Journal of Medicine and Philosophy* 5 (2000): 285–307.

Rollin, Betty. 2000 [1976]. *First, You Cry*. New York: Quill.

Rosenberg, Charles E. 1992. "Introduction." In *Framing Disease: Studies in Cultural History*, edited by C.E. Rosenberg and J. Golden, xiii–xxvi. New Brunswick, NJ: Rutgers University Press.

Ross, W.D. 2011. "What Makes Right Acts Right?" In *Ethical Theory*, 6th ed., edited by L. P. Pojman and J. Feiser, 319–327. Belmont, CA: Wadsworth.

Rosser, Sue V. 2000. "Controversies in Breast Cancer Research." In *Breast Cancer: Society Shapes an Epidemic*, ed. A. S. Kasper and S. J. Ferguson, 245–270. New York: St. Martin's Press.

Sackett, David. L. 2005. *Evidence-Based Medicine*. New York: John Wiley and Sons.

Salz, Talya. 2013. "Survivorship Care Plans in Research and Practice." *CA: A Cancer Journal for Clinicians.* http://www.ncbi.nlm.nih.gov/pmc/articles/pmc3330140 (accessed August 1, 2014).

Sassower, Raphael, and Mary Ann Cutter. 2007. *Ethical Choices in Contemporary Medicine.* Stocksfield: Acumen.

Sassower, Raphael, and Michael Grodin. 1987. "Scientific Uncertainty and Medical Responsibility." *Theoretical Medicine* 8: 221–234.

Schattner, Elaine. 2015. "Study Highlights Social Factors Affecting Breast Cancer Surgery." *Forbes,* June 19. http://www.forbes.com/sites/elaineschattner/2015/06/19/study-highlights-social-factors-affecting-breast-cancer-surgery-decisions-lumpectomy-vs-mastectomy/ (accessed July 1, 2015).

Selleck, Meredith, and Amy Tiersten. 2004. "The Difference Between Male and Female Breast Cancer." In *Principles of Gender-Specific Medicine*. Vol. 2., edited by Marianne Legato, 648–657. Boston: Elsevier Academic Press.

Sherwin, Susan. 1992. *No Longer Patient: Feminist Ethics and Health Care.* Philadelphia: Temple University Press.

Sherwin, Susan. 1998. "A Relational Approach to Autonomy in Health Care." In *The Politics of Women's Health,* edited by S. Sherwin et al., 19–47. Philadelphia: Temple University Press.

Sherwin, Susan. 2006. "Personalizing the Political." In *The Voice of Breast Cancer in Medicine and Bioethics,* edited by M. Rawlinson and S. Lundeed, 3–19. New York: Springer.

Simon, H. 1955. "A Behavioral Model of Rational Choice," *Quarterly Journal of Economics* 69: 99–118.

Skloot, Rebecca. 2010. *The Immortal Life of Henrietta Lacks.* New York: Crown.

Sontag, Susan. 1978. *Illness as Metaphor.* New York: Farrar, Straus, and Giroux.

Starr, Paul. 1982. *The Social Transformation of American Medicine: The Rise of the Sovereign Profession and the Making of a Vast Industry.* New York: Basic Books.

Stempsey, William E. 2004. "The Philosophy of Medicine: Development of a Discipline." *Medicine, Health Care, and Philosophy* 7: 243–251.

Susan G. Komen. 2014a. "International Community Health Grants." http://ww5.komen.org/ResearchGrants/InternationalCommunityHealthGrants.html (accessed July 25, 2014).

Susan G. Komen. 2014b. "Molecular Subtypes of Breast Cancer." http://www5.komen.org/molecular subtypes (accessed August 17, 2014).

Susan G. Komen. 2014c. "Survivorship Topics." http://www5.komen.org/breast-cancer (accessed August 24, 2014).

Sydenham, Thomas. 1753. *The Entire Works of Dr. Thomas Sydenham.* 3rd ed. Edited and translated by John Swan. London: E. Cave.

Taghian, Alphonese G. et al. 2009. *Breast Cancer: A Multidisciplinary Approach to Diagnosis and Management.* New York: Demos Medical Publishing.

Taylor, Charles. 1979. "Atomism." In *Powers, Possessions, and Freedom,* edited by A. Kontos, 39–62. Toronto: University of Toronto Press.

Thagard, Paul. 1999. *How Scientists Explain Disease.* Princeton, NJ: Princeton University Press.

Thibodeau, Gary A., and Kevin T. Patton. 2005. *The Human Body in Health and Disease.* Amsterdam: Elsevier.

Tong, Rosemarie. 1997. *Feminist Approaches to Bioethics: Theoretical Reflections and Practical Applications.* Boulder, CO: Westview.

Tong, Rosemarie. 2014. *Feminist Thought: A More Comprehensive Introduction.* 4th ed. Boulder, CO: Westview.

Tuana, Nancy. 2006. The Spectrum of Ignorance: The Women's Health Movement and Epistemologies of Ignorance." *Hypatia* 21 (3): 1–19.

"Types of Breast Cancer." 2013. February 25. http://www.breastcancer.org/symptoms/types (accessed February 25, 2014).

US Department of Health and Human Services. 2013. "State Laws Relating to Breast Cancer." http://www.cdc.gov/cancer/breast/pdf/BCLaws.PDF (accessed December 7, 2013).

US Department of Labor. 2013. "Your Rights After a Mastectomy." April 7. http://www.dol.gov/ebsa/publications/whcra.html (accessed August 22, 2013).

US Food and Drug Administration. 2015. "Risks of Breast Implants." http://www.fda.gov/MedicalDevices/ProductsandMedicalProcedures/ImplantsandProsthetics/BreastImplants/ucm064106.htm (accessed June 1, 2015).

US Preventive Services Task Force. 2009. "Breast Cancer Screening." http://www.uspreventiveservicestaskforce.org/Page/Topic/recommendation-summary/breast-cancer-screening (accessed May 1, 2015).

Van Epps, Heather L. 2012. "The Estrogen Effect." *Cure* (Fall): 44–49.

Veatch, Robert M. 1981. "The Medical Model: Its Nature and Problems." In *Concepts of Health and Disease: Interdisciplinary Perspectives,* edited by A. L. Caplan et al., 523–544. Boston: Addison-Wesley.

Virchow, Rudolf. 1958 (1895). *Disease, Life, and Man: Selected Essays by Rudolf Virchow*. Trans. L. J. Rather. Stanford, CA: Stanford University Press.

Warnock, G. J. 1967. *Contemporary Moral Philosophy*. New York: St. Martin's Press.

Wartofsky, Marx. 1976. "The Mind's Eye and the Hand's Brain: Toward an Historical Epistemology of Medicine." In *Science, Ethics, and Medicine,* edited by H. T. Engelhardt, Jr. and D. Callahan, 167–194. New York: The Hastings Center.

Watson, James, and Francis Crick. 1953. "A Structure for Deoxyribose Nucleic Acid." *Nature* 171: 737–738.

Weber, Lynn. 2006. "Reconstructing the Landscape of Health Disparities Research: Promoting Dialogue and Collaboration between Feminist Intersectional and Biomedical Paradigms." In *Gender, Race, Class, and Health: Intersectional Approaches,* edited by A.L. Schulz and L. Mullings, 21–59. San Francisco: Jossey-Bass.

"What Is Personalized Cancer Medicine?" http://www.geneticmedicine.tv/what-is-personalized-cancer-medicine-cancer-net/ (accessed June 1, 2017).

Wittgenstein, Ludwig. 1963. *Philosophical Investigations.* Trans. G. E. M. Anscombe. Oxford: Basil Blackwell.

Women's Health Initiative. 2015. "Questions and Answers About the WHI Postmenopausal Hormone Therapy Trials." http://www.nhlbi.nih.gov/whi-faq.htm (accessed June 20, 2015).

World Health Organization. 2013. "International Classification of Diseases Information Sheet." *Classifications*. February 13. http://www.who.int/classification/icd/factsheet/en/index.html (accessed May 24, 2015).

World Health Organization. 2014. "Breast Cancer: Prevention and Control." http://www.who.int/cancer/detection/breastcancer/en/index1.html (accessed August 15, 2014).

World Health Organization. 2015. "The Breast Cancer Conundrum." *Bulletin of the World Health Organization.* http://www.who.int/bulletin/volumes/91/9/13-020913 (accessed May 20, 2015).

Wulff, Henrik R. 1981. *Rational Diagnosis and Treatment: An Introduction to Clinical Decision-making*. 2nd ed. Oxford: Blackwell Scientific Publications.

Wulff, Henrik R. et al. 1986. *Philosophy of Medicine: An Introduction*. Oxford: Blackwell Scientific Publications.

Youlden, Danny R. et al. 2012. "The Descriptive Epidemiology of Female Breast Cancer: An International Comparison of Screening, Incidence, Survival, and Mortality. *Cancer Epidemiology.* March 14. http://www.cancerepidemiology.net/article/S1877-7821%2812%2900029-X/abstract (accessed July 3, 2015).

Yoxen, Edward. 1983. *The Gene Business: Who Should Control Biotechnology?* New York: Free Association Books.

Zola, Irving Kenneth. 1972. "Medicine as an Institution of Social Control." *Sociological Review* 20: 487–504.

INDEX

access to breast cancer care, 171–177
accollaboration, 106–107
accommodification, 119–121
amedicalization, 113–115
antirealism, 25–28
apoliticalization, s124–125
Aristotle, 44, 171
Aronowitz, Robert, xiv, 107, 118
autonomy, 152–161
Ayer, A. J., 83

Bayesian analysis, 58–59
Beauchamp, Tom L., 153–180
beneficence, 164–171
Berkeley, George, 37
biomarkers, 30–32
breast cancer
  case study of, 12–15
  description of, 18–46, 130–134, 186–187
  evaluation of, 79–102, 130–137, 187–188
  explanation of, 47–78, 130–137, 187–188
  genetic account of, 32–33
  global information on, 2, 6–9
  grading of, 50
  history of, 8–10
  immunological account of, 33–34
  laws concerning, 4–8
  nosology of, 1–2
  policies concerning, 4–8
  socialization of, 103–128, 130, 134–137, 188–189
  stages of, 50–53, 82
  statistics on, 2–4
biomedical ethics, 151–184
Boorse, Christopher, 84
Breast Cancer Awareness Month, 127

causal relations, 60–66
certainty
  cognitive, 66–73
  value, 95–100
Childress, James F., 153–180
collaboration, 107–113
Collins, Francis, 147–148
commodification, 121–123
Conrad, Peter, 115
contextual medical phenomenon, 144–145
Crenshaw, Kimberlé Williams, 143
Cutter, Mary Ann, 87, 89, 139–142

Davis, Kathy, 122
Dawkins, Richard, 41
Descartes, René, 57
description of breast cancer, 18–46, 130–134, 186–187
Diogenes Laertius, 74–75
disease, four dimensions of, 15

INDEX

empiricism, 49–56
Engel, George, 46
Engelhardt, H. Tristram, Jr, 13, 14, 23, 73, 85, 99–100, 108, 130–131, 195
Ericksen, Julia, 75, 98
ethics of breast cancer, 11, 151–184
evaluation of breast cancer, 79–102, 130–137, 188–189
explanation of breast cancer, 47–78, 130–137, 187–188

Fleck, Ludwik, 110–111
Foot, Philippa, 87
Foucault, Michel, 59–60
funding for breast cancer, 4

genetic account of breast cancer, 32–33
Gigerenzer, Gerd, 156–158, 168–170
global information on breast cancer, 2, 6–9
Gorgias Leoritini, 74
Gøtzsche, Jerome, 44, 54–56, 64, 113
grading of breast cancer, 50
Grodin, Michael, 67–69
Groopman. Jerome, 58–59, 70–72, 76

history of breast cancer, 8–10
Hobbes, Thomas, 34
holism, 44–46
Hudis, Clifford, 175
Hume, David, 36

idealism, 37–39
immunological account of breast cancer, 33–34
informed consent, 152–161, 190
integrative medical phenomenon, 152–161, 190
intersectional medical phenomenon, 142–144

justice, 171–177, 191

Khushf, George, 92
Kuhn, Thomas, 110

laws concerning breast cancer, 4–8
Lerner, Barron, 112
Locke, John, 54
logical positivists, 124

materialism, 28–35
medical thinking, 114–115
medicalization, 115–118
Morgan, Kathryn Pauly, 120

nonmaleficence, 163–177
normativism
  disease, 86–88
nosology of breast cancer, 1–2

ontological issues, 19
ontological view of disease, 23

Paracelsus, 22
personalized medical phenomenon, 146–149
philosophy of medicine, 12–15
physicalism, 23–35
physiological view of disease, 27–28
Plato, 22
policies concerning breast cancer, 4–8
politicalization, 125–127
principles of biomedical ethics, 153–180
probabilistic thinking, 59–60
Proctor, Robert, 126–127
Protagoras of Abdera, 98

rationalism, 57–60
realism, 19–23
reductionism, 39–44
Reznek, Lawrie, 14
risk assessment, 161–171, 190–191
Romberg, Heinrich, 146–147
Ross, W. D., 96

Sassower, Raphael, 67–69, 139–141
Sherwin, Susan, 159–161
skepticism
  cognitive, 73–77
  value, 100–101
socialization of breast cancer, 103–128, 130, 134–137, 188–189
stages of breast cancer, 50–53, 82
statistics on breast cancer, 2–4

Tamoxifen, 127
Tong, Rosemarie, 181–182
Tuana, Nancy, 193

Uncertainty, managing, 66–77, 95–101, 192–196

value neutralism, 81–86
values
  aesthetic, 92
  clinical, 88–95
  ethical or moral, 93
  fundamental, 89–90
  instrumental, 91
  moral or ethical, 93
Veatch, Robert, 114
Virchow, Rudolf, 26

Wulff, Henrik, 45, 62

Youlden, Danny, 174

Zola, Irving Kenneth, 117

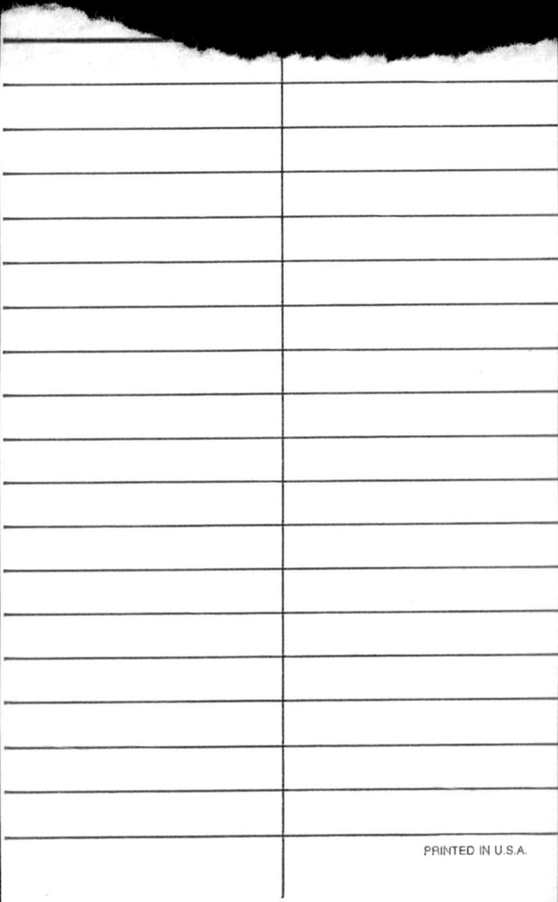
PRINTED IN U.S.A.